how to

how to

absurd scientific advice
for common real-world problems

RANDALL MUNROE

RIVERHEAD BOOKS

NEW YORK

2019

RIVERHEAD BOOKS
An imprint of Penguin Random House LLC

Copyright © 2019 by xkcd inc.

ISBN 9780525537090 (hardcover)
ISBN 9780525537106 (ebook)
ISBN 9780593086377 (international)

Printed in the United States of America

1 3 5 7 9 10 8 6 4 2

Book design by Christina Gleason

Contents

Disclaimer

Do not try any of this at home. The author of this book is an internet cartoonist, not a health or safety expert. He likes it when things catch fire or explode, which means he does not have your best interests in mind. The publisher and the author disclaim responsibility for any adverse effects resulting, directly or indirectly, from information contained in this book.

Hello! ←

This is a book of bad ideas.

At least, most of them are bad ideas. It's possible some good ones slipped through the cracks. If so, I apologize.

Some ideas that sound ridiculous turn out to be revolutionary. Smearing mold on an infected cut sounds like a terrible idea, but the discovery of penicillin showed that it could be a miracle cure. On the other hand, the world is full of disgusting stuff that you *could* smear on a wound, and most of them won't make it better. Not all ridiculous ideas are good. So how do we tell the good ideas from the bad?

We can try them and see what happens. But sometimes, we can use math, research, and things we already know to work out what will happen if we do.

When NASA was planning to send its car-size *Curiosity* rover to Mars, they had to figure out how to land it gently on the surface. Previous rovers had landed using parachutes and air bags, so NASA engineers considered this approach with *Curiosity*, but the rover was too large and heavy for parachutes to slow it down enough in Mars's thin atmosphere. They also thought about mounting rockets on the rover to let it hover and touch down gently, but the exhaust would create dust clouds that would obscure the surface and make it hard to land safely.

Eventually, they came up with the idea of a "sky crane" – a vehicle that would hover high above the surface using rockets while lowering *Curiosity* to the ground on a long tether. This sounded like a ridiculous idea, but every other idea they could come up with was worse. The more they looked at the sky crane idea, the more plausible it seemed. So they tried it, and it worked.

We all start out life not knowing how to do things. If we're lucky, when we need to do something, we can find someone to show us how. But sometimes, we have to come up with a way to do it ourselves. This means thinking of ideas and then trying to decide whether they're good or not.

This book explores unusual approaches to common tasks, and looks at what would happen to you if you tried them. Figuring out why they would or wouldn't work can be fun and informative and sometimes lead you to surprising places. Maybe an idea is bad, but figuring out exactly *why* it's a bad idea can teach you a lot—and might help you think of a better approach.

And even if you already know the right way to do all these things, it can be helpful to try to look at the world through the eyes of someone who doesn't. After all, for anything that "everyone knows" by the time they reach adulthood, every day over 10,000 people in the United States alone are learning it for the first time.

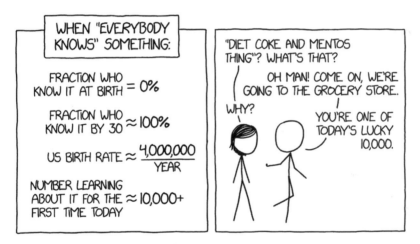

That's why I don't like making fun of people for admitting they don't know something or never learned how to do something. Because if you do that, all it does is teach them not to tell you when they're learning something... and you miss out on the fun.

This book may not teach you how to throw a ball, how to ski, or how to move. But I hope you learn something from it. If you do, you're one of today's lucky 10,000.

how to

CHAPTER 1

How to Jump Really High

People can't jump very high.

Basketball players make some impressive leaps to reach hoops placed high in the air, but most of their reach comes from their height. An average professional basketball player can only jump a little more than 2 feet straight up. Non-athletes are more likely limited to jumping a foot or so. If you want to jump higher than that, you'll need some help.

Using a running start can help. This is what athletes competing in the high jump do, and the world record height is 8 feet. However, that's measured from the ground. Since high jumpers tend to be tall, their center of mass starts off several feet off the ground, and because of how they fold their bodies to pass over the bar, their center of mass may actually pass *under* it. An 8-foot high jump doesn't involve launching the center of their body anything like the full 8 feet.

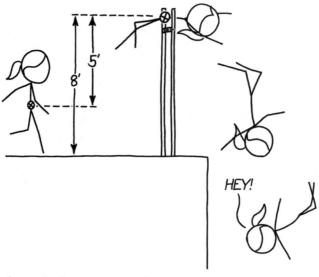

If you want to beat a high jumper, you have two options:

1. Dedicate your life to athletic training, from an early age, until you become the world's best high jumper.
2. Cheat.

The first option is no doubt an admirable one, but if that's your choice, then you're reading the wrong book. Let's talk about option two.

There are a lot of ways you could cheat at high jump. You could use a ladder to get over the bar, but that's hardly *jumping*. You could try wearing those spring-loaded stilts[1] popular with extreme sports enthusiasts, which—if you're athletic enough—might be enough to give you the edge over an unassisted high jumper. But for sheer vertical height, athletes have already come up with a better technique: pole vaulting.

---

1 Or alternatively, for the nineties kids out there, Nickelodeon® Moon Shoes®™

HOW POLE VAULTING WORKS

TYPE 1: STANDARD

TYPE 2: EXTREME

In pole vaulting, athletes start running, stick a flexible pole into the ground in front of them, and launch themselves into the air. Pole vaulters can fling themselves several times higher than the best unassisted high jumpers.

The physics of pole vaulting are interesting, and don't revolve around the pole nearly as much as you might think. The key to vaulting isn't the springiness of the pole, it's the athlete's running speed. The pole is just an efficient way to redirect that speed upward. In theory, the vaulter could use some other method to change direction from *forward* to *up*. Instead of sticking a pole in the ground, they could jump onto a skateboard, go up a smooth curved ramp, and reach just about exactly the same height as the vaulter.

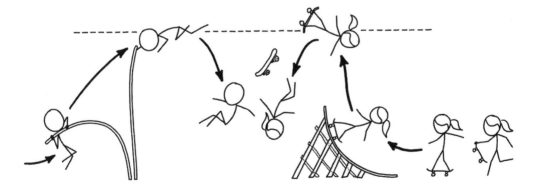

We can estimate a pole vaulter's maximum height using simple physics. A champion sprinter can run 100 meters in 10 seconds. If an object is launched upward at that speed under Earth's gravity, a little math can tell us how high it should go:

$$\text{height} = \frac{\text{speed}^2}{2 \times \text{acceleration of gravity}} = \frac{\left(\frac{100\,\text{meters}}{10\,\text{seconds}}\right)}{2 \times 9.805\,\frac{m}{s^2}} = 5.10\,\text{meters}$$

Since the pole vaulter is running before they jump, their center of gravity starts off above the ground already, which adds to the final height it reaches. A normal adult's center of gravity is somewhere in their abdomen, usually at a height of about 55 percent of their actual height. Renaud Lavillenie, the world record holder in the men's pole vault, is 1.77 meters tall, so his center of gravity adds another 0.97 meters or so, giving a final predicted height of 6.08 meters.

How does our prediction compare to reality? Well, the actual world record height is 6.16 meters. That's pretty close for a quick approximation![2]

Of course, if you show up at a high jump championship with a vaulting pole, you'll be immediately disqualified.[3] But while the judges might object, they probably won't stop you, especially if you wave the pole around threateningly as you approach.

---

2   Physics offers another interesting piece of trivia about pole vault world records. The downward pull of Earth's gravity varies from place to place, both because the shape of the Earth affects its gravity and because the spinning motion "flings" things outward. These effects are small in the grand scheme of things, but the variation from place to place can be as much as 0.7%. That's not enough to notice when you're walking around, but it's enough so that when you buy a scale, you may need to calibrate it, since gravity at the factory could be slightly different from gravity at your house.
The varying pull of gravity is enough to affect pole vault records. In June 2004, Yelena Isinbayeva set the then women's pole vault world record, with a height of 4.87m. She set this record in Gateshead, England. A week later, Svetlana Feofanova broke her record by 1 cm, jumping 4.88m. However, Feofanova set this record in Heraklion, Greece, where the pull of gravity is slightly weaker. The difference is just enough that, if she wanted, Isinbayeva could argue that Feofanova only broke the record because of the weaker gravity, and her Gateshead jump was the more impressive one.
Isinbayeva apparently decided not to make this complicated physics argument, and instead went for a simpler response: a few weeks later, she broke Feofanova's record, again jumping in the stronger British gravity. As of 2017, she still holds the women's record.

3   At least, I assume you would. It's possible no one's ever tried.

IF YOU DON'T *EXACTLY* FOLLOW THE RULES, BUT YOU *SORT* OF FOLLOW THEM, IT'S NOT *TECHNICALLY* CHEATING, RIGHT?

I...WHAT...DO YOU... DO YOU UNDERSTAND WHAT "TECHNICALLY" MEANS?

TECHNICALLY, NO.

Your record won't go on the books, but that's ok—you'll know in your heart how high you jumped.

But if you're willing to cheat more blatantly, you can go higher than 6 meters. A *lot* higher. You just need to find the right spot to launch from.

Runners take advantage of aerodynamics. They wear sleek, tight-fitting outfits to cut down on air resistance, which helps them to gain greater speeds and thus soar higher.[4] Why not take it a step further?

Of course, actually pushing yourself forward with a propeller or a rocket doesn't count. There's no way you can call that a "jump" with a straight face.[5] That's not a jump, that's a *flight.* But there's nothing wrong with just … gliding a little.

The path of every falling object is affected by how the air moves around it. Ski jumpers adjust their shape to gain a huge aerodynamic boost to their jumps. In an area with the right winds, you can do the same thing.

---

4   As of this writing, there's no world record for the highest high jump by an athlete wearing a Victorian hoop skirt, but if there were, it would probably be lower than the regular record.

5   We're cheating, but we're not *cheating.*

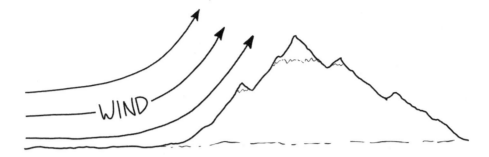

When sprinters run with the wind at their backs, they can reach higher speeds. Similarly, if you jump in an area where the wind is blowing *upward*, you can reach greater heights.

It takes strong wind to push you upward—wind blowing faster than your *terminal velocity*. Your terminal velocity is the maximum speed you'll reach while falling through air, when the force of the air rushing past balances out the downward acceleration of gravity. This is the same as the minimum upward wind speed needed to lift you off the ground. Since all motion is relative, it doesn't really matter whether you're falling downward through the air or the air is blowing upward past you.[6]

People are a lot denser than air, so our terminal velocity is pretty high. A falling person's terminal velocity is around 130 miles per hour (mph). In order to get much of a boost from wind, you'll need the upward wind speed to be at least in the same range as your terminal velocity. If the wind is a lot slower, then it won't affect your jump height very much.

Birds use columns of warm, rising air—called thermals—like elevators. The birds soar in circles without flapping, letting the rising air carry them upward. These thermal updrafts are relatively weak; to lift your larger human body, you'll need to find a stronger source of rising air.

Some of the strongest updrafts near the ground happen near mountain ridges. When wind encounters a mountain or ridge, the airflow can be diverted upward. In some areas, these winds can be quite fast.

Unfortunately, even at the best spots, the vertical winds generally aren't even close to a human's terminal velocity. At best, you'd only gain a little bit of height from the wind's assist.[7]

Instead of trying to increase the wind speed, you can try reducing your terminal velocity with aerodynamic clothing. A good wingsuit—clothing with sheets of material be-

6   At least, from a physics point of view. It probably matters a lot to you personally.

7   You'd also have to convince the judges to hold the competition near the edge of a cliff, which might be difficult.

tween the arms and legs – can reduce a person's sink rate from 130 mph to as little as 30 mph. That's still not enough to actually ride winds upward, but it *would* add some height to your jump. On the other hand, you'd have to do your running approach in a full wingsuit, which would probably cancel out the benefit from the wind.

To add substantial height to your jump, you need to go beyond wingsuits, into the world of parachutes and paragliders. These large contraptions reduce a person's falling speed enough so that surface winds can easily get strong enough to lift them. Skilled paragliders can launch from the ground and ride ridge winds and thermals to thousands of feet.

But if you want a *real* high jump record, you can do even better.

In most areas where air flows over mountains, the "mountain waves" extend up only into the lower atmosphere, which limits the height that gliders can reach. But in some places, when conditions are just right, these disturbances may interact with the polar vortex and polar night jet,[8] creating waves that reach into the stratosphere.

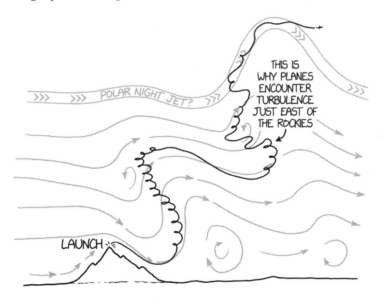

In 2006, glider pilots Steve Fossett and Einar Enevoldson rode stratospheric mountain waves to over 50,000 feet above sea level. That's almost twice as high as Mt. Everest, and higher than the highest commercial airline flights. That flight set a new glider altitude record. Fossett and Enevoldson say they could have ridden the stratospheric waves even

8   The *polar night jet* is a high-altitude wind stream that exists near the Arctic and Antarctic at certain times of year. Not to be confused with *The Polar Night Jet*, a heartwarming children's picture book about a child who visits Santa one night by flying to the North Pole in a magical stealth bomber.

higher – they only turned back because the low air pressure caused their pressure suits to inflate so much that they couldn't operate the controls.

If you want to jump high, you just need to construct an outfit shaped like a sailplane – you can make one out of fiberglass resin and carbon fiber – and head to the mountains of Argentina.

If you find the right spot, and if conditions are *just* right, you can seal yourself into your sailplane suit,[9] jump into the air, catch the ridge lift, and ride the wind into the stratosphere. It's possible that a glider pilot riding these waves might be able to cruise at higher altitudes than any other wing-borne aircraft. That's not bad for a single jump![10]

If you're really lucky, maybe you can find a spot that's upwind from where the Olympics are being held. That way, when you jump off the edge, the winds in the stratosphere will carry you over the venue...

---

9 You'll need to pressurize the sailplane cabin around you, but that shouldn't be too hard, right? Just make the fiberglass shell airtight, but add a hose for breathing. When you get a few miles up and the air pressure really starts dropping, just pinch off the hose to seal yourself in. You might be up there for a while, so try to make the cabin big enough that you won't run out of air.

10 We forgot doors, so when you land, call a friend to come break open your sailplane with a hammer.

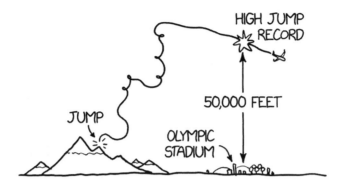

… letting you set the greatest high jump record in the history of the sport.

They probably won't give you a medal, but that's fine. You'll know you're the real champion.

CHAPTER 2

How to Throw a Pool Party

You've decided to throw a pool party. You've got everything—snacks, drinks, floating inflatable toys, towels, and those rings you throw into the pool and then have to dive in to retrieve. But the night before the party, you can't shake the feeling that you're missing something. Looking around your yard, you realize what it is.

You don't have a pool.

Don't panic. You can solve this problem. You just need a bunch of water and a container to put it in. Let's figure out the container first.

There are two main types of pools: *in-ground* and *above-ground*.

IN-GROUND POOL

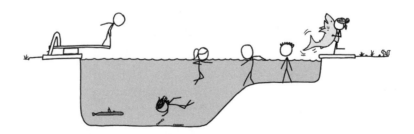

An in-ground pool is, when it comes down to it, a fancy hole. This type of pool can take more work to install, but is also less likely to collapse in the middle of your party.

If you'd like to build an in-ground pool, first consult chapter 3: How to Dig a Hole. Use those instructions to dig a hole roughly 20 feet by 30 feet by 5 feet. Once you've created a hole of the appropriate size, you may want to line the walls with some kind of coating to keep the water from turning to mud or draining out before the party is over. If you have some giant plastic sheets or tarps lying around, you can use those, or you can try a spray-on rubber coating—there are ones designed for lining the beds of koi ponds. Just tell the salespeople you have some really large koi.

ALTERNATIVE: ABOVE-GROUND POOL

If you decide an in-ground pool isn't the option for you, you can instead try an above-ground pool. The design of this type of pool is relatively simple:

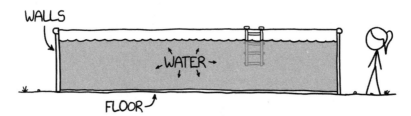

Unfortunately, water is heavy – ask anyone who's ever filled a fish tank on the floor and then tried to lift it up onto a table. Gravity pulls the water downward, but the ground pushes back equally hard. The water pressure is redirected outward, toward the walls of the pool, which are stretched in all directions. This tension, called *hoop stress*, is strongest at the base of the wall where the water pressure is the highest. If the hoop stress exceeds the tensile strength of the wall, the wall will burst.[1]

Let's pick a possible material – say, aluminum foil. How deep can the water in an aluminum-foil-walled pool get before the sides burst? We can figure out the answer to this question, and lots of other pool design questions, using the formula for hoop stress:

$$\text{hoop stress} = \text{water depth} \times \text{water density} \times \text{Earth gravity} \times \frac{\text{pool radius}}{\text{wall thickness}}$$

---

1   In practice, it will probably burst before that point, thanks to irregularities in the materials and their particular "yield curves," but we can use simple tensile strength as an approximation.

Let's plug in the numbers for aluminum foil. Aluminum has a tensile strength of around 300 megapascals (MPa), and sheets of foil are roughly 0.02 mm thick. Let's assume our pool is 30 feet in diameter, so there's plenty of room for games. We can plug those values into the hoop stress equation and rearrange things to figure out how deep the water in our shiny, crinkly pool can get before the hoop stress equals the tensile strength of the aluminum and the walls give out:

$$\text{water depth} = \frac{\text{wall thickness} \times \text{wall tensile strength}}{\text{water density} \times \text{gravity} \times \text{pool radius}}$$

$$= \frac{0.02 \text{ mm} \times 300 \text{ MPa}}{1 \frac{\text{kg}}{\text{L}} \times 9.8 \frac{\text{m}}{\text{s}^2} \times \frac{30 \text{ feet}}{2}} \approx 5 \text{ inches}$$

Sadly, 5 inches of water is probably not enough for a pool party.

If we swap out the thin aluminum foil for inch-thick pieces of wood, the math looks much better. Wood has a lower tensile strength than aluminum foil, but it makes up for it by being thicker, and could hold water 75 feet deep. If you happen to have a 30-foot-wide wooden cylinder with inch-thick walls lying around, that's great news for you!

You can also rearrange the equation to tell you how thick the pool's walls need to be to support a desired water depth. Let's say we want our pool to be about 3 feet deep. Given the tensile strength of a material, this version of the formula tells us the minimum wall thickness necessary to hold the water:

$$\text{wall thickness} = \frac{\text{water depth} \times \text{water density} \times \text{gravity} \times \text{pool radius}}{\text{wall tensile strength}}$$

The great thing about physics is that you can run these numbers for any material you want, even if it's something ridiculous. Physics doesn't care if your question is weird. It just gives you the answer, without judging. For example, according to the comprehensive 456-page handbook *Cheese Rheology and Texture*, hard Gruyere cheese has a tensile strength of 70 kPa. Let's plug that into the formula!

$$\text{wall thickness} = \frac{3 \text{ feet} \times 1\frac{\text{kg}}{\text{L}} \times 9.8\frac{\text{m}}{\text{s}^2} \times \frac{30 \text{ feet}}{2}}{70 \text{ kPa}} \approx 2 \text{ feet}$$

Good news! You'll only need a two-foot-thick wall of cheese to contain your pool! The bad news is that you may have trouble convincing anyone to jump in.

Given the practical problems associated with cheese, you should probably stick to traditional materials like plastic and fiberglass. Fiberglass has a tensile strength around 150 MPa, which means a wall just a millimeter thick would be strong enough to contain the water with extra tensile strength to spare.

GET SOME WATER

Now that you've got your pool — whether in-ground or above-ground — you'll need some water. But how much?

Standard backyard in-ground pools vary in size, but a medium-size one large enough to have a diving board might hold 20,000 gallons of water.

If you have a garden hose and a municipal water supply, then you could potentially fill your pool that way. But whether or not you can fill a pool *quickly* depends on the flow rate from your hose.

If you have good water pressure and a large-diameter hose, your flow rate might be 10 or 20 gallons per minute, which is enough to fill your pool within a day or so. If your flow rate is too low — or if you have well water, which may run out before you fill your pool — you might need to look for a different solution.

INTERNET WATER

In many areas, online retailers like Amazon offer same-day delivery. A 24-pack of Fiji water bottles currently costs about $25. If you have $150,000 to spare – plus another $100,000 or so for same-day delivery – you can simply order a pool in bottle form. As a bonus, your new pool will consist entirely of water shipped from Fiji.

This will present a new challenge. When the water is delivered, you'll need to get it all into the pool.

This will be trickier than you might have thought. Sure, you could unscrew the cap on each bottle and dump the water into the pool one by one, but this would take a few seconds per bottle. Since there are 150,000 bottles and only 86,400 seconds in a day, anything that takes more than a second per bottle is definitely not going to work.

ATTACK THE BOTTLES

You could try slicing the caps off an entire 24-pack of bottles with a sword. Many slow-motion videos online show people cutting through a row of water bottles with a sword. Judging from the videos, it's surprisingly hard to do – the sword tends to be deflected up or down as it passes through the bottles. Even if you had a precise enough swing, along with the requisite arm strength and endurance, using a sword would probably be too slow.

Guns probably wouldn't work too well, either. With careful planning and an efficient setup, you could use some kind of shotgun to make holes in a whole case of bottles at once, but you'd still have a hard time puncturing all the bottles and making them all fully drain fast enough to get through them all. You'd also end up with a pool full of lead, which—especially if you added chlorine to the water—would corrode and could eventually contaminate the groundwater.

There are a wide variety of increasingly powerful weapons you could use to try to open these bottles quickly; we won't run through them all here. But before we leave weapons behind and move on to a more practical solution, let's take a moment to consider the biggest and most impractical option of all. Could you open bottles using nuclear bombs?

This is a completely ridiculous suggestion, so it should come as no surprise that it was studied by the US government during the Cold War. Early in 1955, the Federal Civil Defense Administration bought beer, soda, and carbonated water from local stores, then tested nuclear weapons on them.[2]

Now, they weren't trying to *open* the beverages. The purpose of the test was to see how well the containers survived, and whether the contents were contaminated. The civil defense planners figured that, after a nuclear explosion in a US city, potable water would likely be needed by first responders, and they wanted to know whether commercial beverages would be a safe source of hydration.[3]

The saga of the government's nuclear war on beer is cataloged in a 17-page report titled *The Effect of Nuclear Explosions on Commercially Packaged Beverages*, a copy of which was helpfully unearthed by nuclear historian Alex Wellerstein.

---

2   The drinks, not the stores.

3   They focus in particular on beer, which seems like it would be less than ideal in a post-nuclear-attack recovery scenario, and it makes you wonder whether the entire test program was hastily arranged as a cover story when someone was caught charging drinks to their work account.

The report describes how the bottles and cans were placed in various locations around the Nevada test site for each explosion. Some were in refrigerators, some on shelves, and some just sitting on the ground.[4] They carried out the experiment twice, during two different nuclear tests conducted as part of Operation Teapot.

The beverages fared surprisingly well. Most of them survived the blast intact. Those that didn't were mostly punctured by flying debris or exploded when knocked off the shelves. They also had low levels of radioactive contamination, and even tasted ok.

Post-explosion beer samples were sent for "carefully controlled testing" at "five qualified laboratories."[5] The consensus was that the beer mostly tasted fine. They concluded that beer recovered after a nuclear blast could be considered a safe source of emergency hydration, but that it should probably be tested more carefully before it was put back on the market.

Plastic bottles weren't common in the 1950s, so all the tests used glass and metal bottles. However, the tests still suggest that nuclear weapons probably don't make great bottle openers.

INDUSTRIAL SHREDDERS

Luckily for us, there's a type of device which can accomplish our goal much more quickly than a sword, shotgun, or nuclear weapon: an industrial plastic shredder. Shredders are used by recycling centers to shred large volumes of plastic bottles, and – as a bonus – they can strain out the liquid for you.

4   In an example of oddly excessive attention to detail, the bottles on the ground were placed at a variety of carefully measured angles relative to ground zero – some lying with the top or bottom end pointed toward ground zero, some lying at 45° angles, and some upright. Perhaps they wanted to see which way you should store your bottles relative to the city center to maximize the chance they'd survive a nuclear attack.

5   I hope this is a euphemism for "friends of ours."

A shredder like the Brentwood AZ15WL 15kW can handle a throughput of 30 tons per hour—including both plastic and liquid, according to Brentwood marketing materials. This would let you fill your pool in a little over 2 hours.

Industrial shredders come with price tags in the five to six figures, which is a lot for one party (although it's nothing compared to what you already spent on water bottles.) But maybe, if you mention how many nuclear weapons you have, you can convince them to give you a discount.

LET SOMEONE ELSE DO THE WORK

If someone else has a pool nearby, and they're at a slightly higher elevation, you can steal the water using a siphon. If you can connect the two pools with a tube of water, you can get water to flow steadily from their pool into yours.

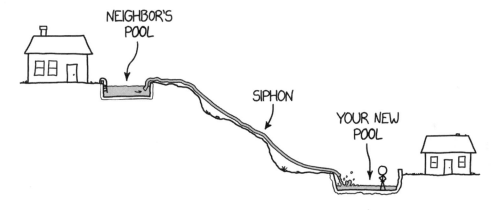

Note: Siphons can lift water up out of a pool and over small barriers like fences, but if the middle of the siphon goes more than 30 feet above the surface of your neighbor's pool, water won't flow. Siphons are driven by atmospheric pressure, and Earth's air pressure is only capable of pushing water up 30 feet against gravity.

GET WATER BY MAKING IT

Water is made up of hydrogen and oxygen. There's plenty of oxygen in the atmosphere,[6] and while hydrogen is certainly rarer, it's still not too hard to find.

The good news is that if you get a bunch of hydrogen and oxygen together, it's easy to turn it into water. You just apply a little bit of heat, and the chemical reaction keeps going. In fact, it's pretty hard to stop.

I'VE FIGURED OUT A WAY TO PRODUCE THE OXIDATION REACTION WE NEED, AND IT LOOKS LIKE IT'S ACTUALLY SELF-SUSTAINING!

FIRE. YOU'RE DESCRIBING FIRE.

The bad news is that sometimes the chemical reaction gets started by accident. We used to have big hydrogen-filled airships flying around, but after some dramatic incidents in the 1930s, we started filling them with helium instead. Nowadays, if you want hydrogen, the best place to get it is by collecting and reprocessing the byproduct of fossil fuel extraction.

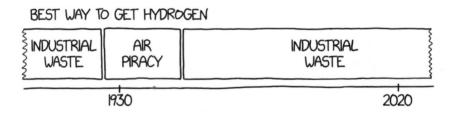

BEST WAY TO GET HYDROGEN

INDUSTRIAL WASTE	AIR PIRACY	INDUSTRIAL WASTE

1930 — 2020

---

6 As of 2019.

GET WATER FROM THE AIR

You don't need to combine hydrogen and oxygen to create water when there's already-created H_2O floating around in the air in the form of water vapor—the stuff that condenses to form clouds and sometimes even falls in the form of rain. On average, each square meter of the Earth has about 6 gallons of water in the column of air above it, the equivalent of a couple of 24-packs of water bottles.[7]

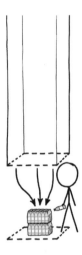

If all that water fell as rain, it would form a layer about an inch thick. If your property is 1 acre in size, and the air has an average amount of moisture, then there are about 25,000 gallons of water in the air overhead. That's enough to fill a pool! Unfortunately, a lot of that water is pretty high up and hard to get to. It would be nice if we could make the water fall on cue, but despite periodic attempted cloud-seeding projects, no one has found a way to reliably induce rainfall.

---

7   This is just an average—the total amount of water per square meter varies from almost nothing, in cold air over deserts, to up to 20 gallons per square meter of land on a humid day in the tropics.

The usual way of extracting water from air is to make the air flow past a cold surface, so the water condenses out of it as dew. To get all the water out of your air, you'd need to build a several-mile-high cooling tower. Luckily for you, air moves around on its own, so you don't need to build a mile-high tower—if there's a breeze, you can just collect the moisture from the air as it flows past your house.

Moisture collection is really a pretty inefficient way to gather water. It takes a lot of power to cool and condense water out of the air. In most cases, you'd use a lot less energy by just driving a truck to an area with more water, filling it up, and driving back. Besides, even under ideal conditions, this kind of humidifier is unlikely to produce enough water to fill your pool any time soon, and it might annoy your neighbors who live downwind of you.

WHY DOES MY SKIN FEEL SO DRY ALL OF A SUDDEN?

GET WATER FROM THE SEA

There's a lot of water in the sea,[8] so probably no one will mind if you borrow a little. If your pool is below sea level, and you don't mind a saltwater pool, this might be an option. All you need to do is dig a channel and let the sea flow in.

This has actually happened in real life, by accident, very dramatically.

Malaysia was once the world's largest producer of tin. One of the mines that produced this tin was constructed near the western coast, just a few hundred feet from the ocean. After the tin market collapsed in the 1980s, the mine was abandoned. On October 21, 1993, the water broke through the narrow barrier separating the mine from the sea. The ocean rushed in, filling the mine in a matter of minutes. The lagoon created by the flood remains to this day, and can be seen on maps at 4.40°N, 100.59°E. The cataclysm was recorded by a bystander with a camcorder, and the footage has since been uploaded to the internet. Despite its low quality, it's one of the most jaw-dropping pieces of video ever recorded.[9]

If the bottom of your pool is above sea level, connecting it to the ocean won't work; water would just flow downhill to the sea. But what if you could bring the sea up to *you*?

Well, you're in luck; it's happening whether you want it to or not. Thanks to the trapped heat caused by greenhouse gases, the seas have been rising for many decades now. Sea-level-rise is caused by a combination of melting ice and thermal expansion of the water. If you want to fill your pool, you could try accelerating sea-level rise. Sure, it would worsen the immeasurable ecological and human toll of climate change, but on the other hand, you could have a sweet pool party.

---

8   [Citation needed]

9   Search for **Pantai Remis landslide**.

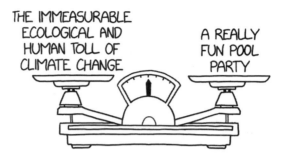

If you wanted to cause rapid sea-level-rise, and you happened to have a giant ice sheet on the land next to your house, you might think that melting it would be a great way to raise the sea level.

But, because of some counterintuitive physics, melting an ice sheet next to your house might actually *lower* the sea level. What you want to do is melt ice on the *other side of the world*.

The reason for this bizarre effect is gravity. Ice is heavy, and when it's sitting on land, it pulls the ocean slightly toward it. When it melts, the water level goes up on average, but since it's no longer being pulled as hard toward the land, it can actually go down in the area around the ice that melted.

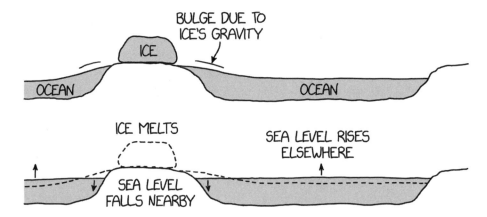

When ice from the Antarctic ice cap melts, sea level goes up the most in the northern hemisphere. When ice in Greenland melts, on the other hand, it raises sea levels most around Australia and New Zealand. If you want to raise sea level near where you live, check whether there's an ice sheet on the other side of the planet. If so, that's the one you should melt.

GET WATER FROM THE LAND

If there are no convenient ice caps to melt—or you don't want to contribute to global sea-level-rise—you could try to do what farmers have been doing to get water for thousands of years: borrow a river.

You could find a nearby river and encourage it—via a temporary dam—to flow toward your pool long enough to fill it up. But be careful: this kind of project has gone wrong before.

In 1905, engineers on the California/Arizona border were digging irrigation canals to bring water to farms from the Colorado River. The mission to divert water from the Colorado River was, unfortunately, *too* successful. The water flowing into the new canal started eroding a deeper and wider path, which let more water flow in. Before they could pull the plug,[10] the river had been captured completely. It inundated a formerly dry valley downstream from the irrigation project, filling it and creating a new—completely accidental—inland sea.

---

10 ▸ Or put in the plug, I guess.

The Salton Sea, which has waxed and waned over the last century, is currently drying up as more water is diverted for irrigation. The windblown dust from the dry lake bed, contaminated with agricultural runoff and other pollutants, blows through nearby towns, sometimes making it hard to breathe. The contaminated, increasingly salty water has led to massive die-offs of aquatic life, and the decaying algae and dead fish have created an omnipresent rotten-egg smell which occasionally wafts west as far as Los Angeles.

That might sound bad, but don't worry – those disastrous environmental consequences took a while to develop.

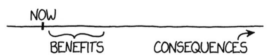

In fact, the Salton Sea was briefly popular as a resort destination, with yacht clubs, fancy hotels, and swimming. Later, as conditions in the sea deteriorated, the resorts all turned to ghost towns. But you can worry about all those consequences tomorrow.

For now, it's pool party time!

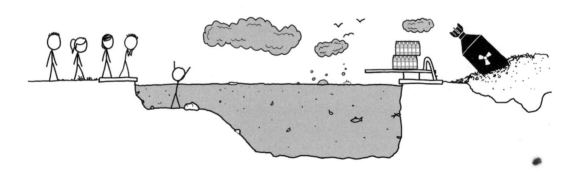

CHAPTER 3

How to Dig a Hole

There are lots of reasons to dig holes. You might be planting a tree, installing an in-ground pool, or putting in a driveway. Or perhaps you've found a treasure map, and you're digging at the *X*.

The best way to dig a hole depends on the size of the hole you want to create. The simplest digging tool is a shovel.

DIGGING WITH A SHOVEL

The rate at which you can dig using a shovel will depend on what kind of dirt you're trying to excavate, but a person digging with a shovel can typically remove between 0.3 and 1.0 cubic meters of dirt per hour. At those rates, in 12 hours, you might be able to excavate a hole about this large:

But if you're digging a hole to get at buried treasure, at some point you may want to stop to consider the economics of the situation.

Digging holes is labor, and labor is valuable. According to the Bureau of Labor Statistics, construction laborers earn an average of $18 per hour. The rate a contractor might charge for an excavation project would also include the cost of planning, equipment, transportation to and from the work site, and disposal of any waste, and likely works out to an hourly rate several times higher. If you spend 10 hours digging a hole in order to find treasure worth $50, you're working for far below minimum wage. In principle, you'd be better off just getting a job digging up driveways somewhere, and in the end, you'd make more money than you would from the treasure.

You also might want to double-check the authenticity of your pirate treasure map, because pirates didn't actually bury treasure.

That's not quite true. There was one time that a pirate buried treasure somewhere. **One time**. And the entire idea of buried pirate treasure comes from that *one* incident.

BURIED PIRATE TREASURE

In 1699, Scottish privateer[1] William Kidd was about to be arrested for various maritime crimes.[2] Before sailing to Boston to confront the authorities, he buried some gold and silver on Gardiners Island, off the tip of Long Island in New York, for safekeeping. It wasn't exactly a secret—he buried it with the permission of the island's proprietor, John Gardiner, along a pathway west of the manor house. Kidd was arrested and eventually executed, and the island's proprietor handed the treasure over to the Crown.

Believe it or not, that's the *entire history of buried pirate treasure*. The reason "buried treasure" is such a well-known trope is that Captain Kidd's story helped inspire Robert Louis Stevenson's novel *Treasure Island*, which almost single-handedly[3] created the modern image of the pirate.

---

1 Pirate.

2 Piracy.

3 Pirates do a lot of things single-handedly.

In other words, this is the only pirate treasure map that has ever existed, and the treasure is gone now:

The scarcity of actual buried pirate treasure hasn't stopped people from searching. After all, just because pirates didn't bury treasure doesn't mean there's never anything valuable in the ground. People who dig a lot of holes, from treasure hunters to archaeologists to construction workers, certainly find valuable stuff from time to time.

But perhaps there's also something compelling about the act of digging for treasure itself—because sometimes people seem to go a little overboard.

OAK ISLAND MONEY PIT

Since at least the mid-1800s, people have believed there's buried treasure near a particular spot on Oak Island in Nova Scotia. Successive groups of treasure hunters have dug deeper and deeper holes in attempts to unearth the treasure. The actual origin of the stories is murky, but at this point it's become almost a meta-myth: most of the evidence that something mysterious is buried on Oak Island consists of stories *about* evidence that may or may not have been found by previous searchers.

> AN OLD FAMILY STORY SAYS MY GRANDFATHER CAME HERE LOOKING FOR BURIED TREASURE.
>
> I'VE FOUND EVIDENCE THAT SOMEONE WAS DIGGING HERE 50 YEARS AGO! IT COULD BE HIM!
>
> MAYBE THE STORIES ARE TRUE!
>
> WE MUST FINISH HIS WORK!

No treasure has been found. Even if a large chest of gold had been buried on the island, the value of the aggregate time and effort that successive generations of treasure hunters have invested in searching for it would by now almost certainly exceed the value of the treasure.

So how big a hole is it worth digging to recover different kinds of treasure?

A single gold doubloon – the classic pirate treasure – is currently[4,5] worth about $300. If you know where a doubloon is buried, it's not worth it to hire someone to dig it out unless the job costs less than $300. And if you value your own labor at $20/hour, then you shouldn't spend more than 15 hours digging it up.

On the other hand, if the treasure is a chest of gold, it could be worth a lot more than $300. A single 1-kilogram gold bar is worth about $40,000, so a chest containing 25 gold bars is worth about a million dollars. If the hole you need to dig is more than 20,000 cubic meters – equivalent to a 30 m × 30 m × 20 m hole – then it will take you so long to dig that the value of the labor involved in the digging will be greater than the value of the treasure. At that rate, you'd be better off just getting a job as a contractor doing excavation work.

The most valuable single piece of traditional "treasure" in the world might be a 12-gram gemstone known as the Pink Star diamond, which sold at auction in 2017 for $71 million. Seventy-one million dollars is enough money to hire a contractor to dig for over a thousand years, or a thousand contractors to dig for over a year. If you owned a 1-acre plot of land, and you knew the Pink Star diamond was buried 1 meter deep somewhere on your property, it would almost certainly be worth the expense to try to dig it up. But if your land were a square kilometer in area and the diamond were buried several meters down, the cost of hiring people to excavate would start to approach $71 million, and digging it up wouldn't be worth it.

---

4   For context, I'm writing this in the year 1731.

5   Note to any far-future historians who found this page and are trying to figure out what year it was really written: That was just a joke. I'm writing this in 2044 CE. from my airship circling the South Pole. I'm so glad this manuscript survived to serve as your Rosetta Stone, and I promise to take this responsibility seriously. By the way: here in the year 2044, we all worship dogs, fear clouds, and eat nothing but honey on the day of a full moon.

At least, it wouldn't be worth it if you were digging with shovels.

VACUUM EXCAVATION

If your planned excavation is large enough that it would take years to dig by hand, then a shovel is almost certainly not the most efficient way to do it, and you should consider slightly more modern techniques.

One more modern digging technique is *vacuum excavation*. Vacuum excavation involves using what is effectively a giant vacuum cleaner to remove the dirt. Suction alone isn't powerful enough to pull apart tightly packed earth, so vacuum excavation combines an industrial vacuum with a jet of high-pressure air or water used to break up the ground.

Vacuum excavation is particularly useful when you want to dig up an area without damaging underground objects such as tree roots, utility lines, or buried treasure. The high-pressure air blows dirt out of the way but leaves larger buried objects intact. Vacuum excavators can remove many cubic meters per hour, potentially expanding the rate at which you can dig by a factor of 10 or more.

The largest holes are dug using mining excavators, which can remove successive layers of land to create *open-pit mines*, holes shaped like an upside-down layer cake. These holes can reach staggering sizes—the Bingham Canyon copper mine in Utah features a central pit about 2 miles across and over half a mile deep.

Oak Island, the site of the infamous money pit, is less than a mile across at its widest point. If the Bingham Canyon excavation had occurred there—with the installation of pumps and seawalls[6] to keep water out of the pit—excavators could have removed the entire island and the underlying bedrock down to a depth 10 times deeper than the deepest shaft dug by treasure hunters.

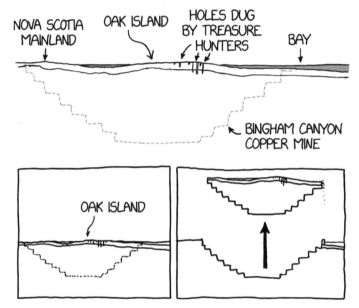

The material could be carefully sifted through to search for any treasure, putting an end to the mystery once and for all.

REMEMBER WHEN THEY DUG UP OAK ISLAND?
MY GRANDFATHER SAID SEVERAL TRUCKLOADS OF
BEDROCK WERE SECRETLY SHIPPED TO CALIFORNIA
AND BURIED THERE WITHOUT BEING SEARCHED.

LET'S GO LOOK FOR THEM!

OH NO.

---

6   A seawall is just a reverse above-ground pool, so you can use the calculation from chapter 2: How to Throw a Pool Party to figure out the engineering. Just use compressive strength instead of tensile strength in the equation.

THE BIGGEST HOLES

Using industrial excavation and drilling methods, humans are capable of digging huge holes. We've removed entire mountains, created vast artificial canyons, and drilled shafts a significant fraction of the way through the Earth's crust. As long as the rock is cool enough to work with, we can dig holes as deep as we want.

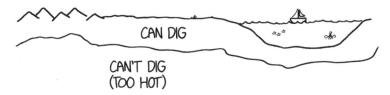

But *should* we?

In 1590, more than 300 years before the Panama Canal was built, the Spanish Jesuit priest José de Acosta discussed the idea of digging a channel through the isthmus to connect the two oceans. In his book *Historia Natural y Moral de las Indias*, he speculated about the potential benefits and pondered some of the engineering challenges involved in "opening the earth and joining the seas." He ultimately decided it was probably a bad idea. Here's his conclusion, from the 2002 translation by Frances López-Morillas:

> *I believe that no human power is capable of tearing down the strong and impenetrable mountain that God placed between the two seas, with hills and rocky crags able to withstand the fury of the seas on either side. And even if it were possible for men to do it I believe it would be very reasonable to expect punishment from Heaven for wishing to improve the works that the Maker, with sublime prudence and forethought, ordered in the fabric of this world.*

Theological questions aside, there's something to be said for his humility. Humans are capable of unlimited excavation, from backyard shovel digs to canal construction to industrial strip mining and mountain removal. And by digging holes, we can certainly find things of value.

But perhaps—sometimes—it's better to just leave the ground the way it is.

CHAPTER 4

How to Play the Piano

(the *whole* piano)[1]

THE PIANO: A DEVICE CAPABLE OF PRODUCING AN INCREDIBLE RANGE OF SOUNDS UNTIL SOMEONE ASKS YOU TO STOP.

---

1   Thank you to Jay Mooney, whose question prompted this chapter.

Playing the piano isn't very hard, in the sense that the keys are all easy to reach and they don't take very much force to push down. Playing a piece of music is just a matter of finding out which keys you need to press, then pressing them at the right time.

Most piano music is written using standard musical notation, which consists of a series of horizontal lines with marks along them corresponding to notes. The higher a note mark is, the higher-pitched the note. Most of the time, notes are drawn in the middle of the lines, but particularly high or low notes sometimes go off the bars to the top or the bottom. A piece of piano music looks something like this:

The standard full-size piano has 88 keys, each of which corresponds to a musical note, organized from lowest on the left to highest on the right. If you see marks above the lines on your sheet music, you'll probably need to press keys on the right side of the piano, whereas marks below the lines will likely mean pressing keys on the left.

A piano can play notes pretty far above and below the lines. In fact, it has one of the widest ranges of any musical instrument, which means it can play all the notes that most other instruments can.[2] If you memorize all the keys and all the notes, then practice playing them in the right order with the right timing, you're all set—you can play any piece of piano music.

PLAYING THE PIANO IS EASY. YOU JUST MEMORIZE WHICH NOTES GO WITH WHICH KEYS, THEN PLAY ALL THE ONES ON THE PAGE IN ORDER.

I'VE EMAILED YOU A LIST OF THE KEYS. THIS CONCLUDES YOUR PIANO LESSON. GIVE ME A CALL IF YOU NEED ANYTHING ELSE. GOOD LUCK!

Well... almost any. The standard full-size piano may have a very wide range, but there are still some notes it can't play. To play *those* notes, you'll need more keys.

When you press a key on a piano, a hammer strikes one or more strings, which vibrate and produce sound. The longer the string is, the lower-pitched the sound will be. Technically, the sound made by each string as it vibrates isn't just a single frequency—it's a rich mix of different frequencies—but each one has a central "main" frequency. The main frequency of the sound made by the leftmost key on a full-size piano is 27 hertz (Hz)—which means the string oscillates 27 times per second—while the rightmost key's main frequency is 4,186 Hz. The ones in between form a regular scale, spanning a range of about 7 octaves. Each key has a frequency roughly 1.059 times higher than the one to its left—that's $2^{1/12}$, which means that every 12 keys, the frequency doubles.

The upper limit of human hearing is quite a bit higher than 4,186 Hz. Young children can hear sounds as high as 20,000 Hz. If we want to be able to play all the notes humans can hear, we'll need to add some keys to the piano. Covering the range between 4,186 Hz and 20,000 Hz requires 27 extra keys.

---

2   Which makes you wonder why we need all those other instruments.

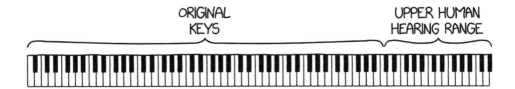

As people get older, they typically lose the ability to hear some of the highest frequencies, so you won't need all the keys if you're playing music for grown-ups. The rightmost handful will produce notes only audible to small children.

On the left end of the piano, covering the human hearing range is a little easier. The lower limit of human hearing is somewhere around 20 Hz, 7 Hz lower than the lowest key on the piano. To cover this range, we'll need another 5 keys. This new, improved 120-key piano will allow you to play any piano music humans are capable of hearing!

But we can extend the piano further.

Sounds above the range of human hearing are called *ultrasound*. Dogs can hear sounds as high as 40 KHz, twice the highest frequency that humans can hear. This is how "dog whistles" work—they produce sounds that dogs can hear but humans can't. Modifying your piano to play dog music will require adding 12 to 15 keys.

Cats, rats, and mice can hear even higher frequencies than dogs, and would need several more keys. Bats—which catch insects by emitting pulses of ultrasound and listening for the echoes—can hear up to around 150 KHz. To cover the full range of hearing for humans, dogs, and bats would require a total of 62 new keys on the right, for a total of 155 keys.

What about even higher frequencies? Unfortunately for us,[3] physics starts to get in the way. High-frequency sounds are absorbed by air as they travel through it, so they fade out quickly. That's why nearby thunder makes a higher-pitched "cracking" sound, while faraway thunder makes a low rumble. They both sound the same at the source, but over a long distance, the high-frequency components of the thunder are muffled and only the low-frequency ones reach your ear.

One hundred fifty kilohertz sound can only travel a few dozen meters in air—which is probably why bats don't use higher frequencies. Since attenuation is related to the square of the frequency, higher ultrasonic pitches are substantially more muffled. If you go too much higher than 150 KHz, the sound won't be able to travel very far beyond your piano. Ultrasonic sounds can travel farther in water or solid material—which is how electric toothbrushes, medical ultrasounds, and high-frequency whale and dolphin echolocation work—but since pianos are typically used in air,[4] 150 KHz makes a pretty good cutoff.

The right side of our piano is complete. What about the left?

Sounds below the normal hearing limit of 20 Hz are called *infrasound*, and they can be a little confusing to think about.

When individual sounds happen quickly enough, they blur into a single hum. Imagine the sound of a bicycle wheel with something stuck in the spokes—at low speeds, it makes a "click click click" sound, while at high speeds it makes a buzzing hum. This suggests that low-frequency sounds shouldn't really "fall below the range of human hearing"—they should just separate into a series of individual sounds. But that isn't quite right.

When sounds are made up of complex individual "pulses"—like the raspy sound of a playing card striking a bicycle spoke—they *do* separate out into individual audible pulses, but only because those pulses are made up of higher-frequency components within normal hearing range. A pure tone, on the other hand, is just a simple sine wave; the sound is made up of air moving smoothly forward and backward. When it slows down to below 20 cycles per second, there are no "clicks" to hear. It just becomes a pulsating pressure wave. We may *feel* it, as a pressure change in the air or a sensation on our skin, but our ears don't interpret it as sound.

Elephants can hear infrasound. Their hearing reaches down to somewhere around 15 Hz—and possibly lower—which means our piano will need at least another 5 keys if we want to play elephant music.

---

3   (But fortunately for our piano tuner)

4   For instructions on how to play the piano underwater, see *How To 2: How to Do a Bunch More Stuff, If You're Still Alive after Following the Instructions in the First Book.*

Sounds below 15 Hz can be detected using specialized equipment. In fact, if you're interested in *very* low frequencies, you can technically make an "infrasound microphone" with just a barometer and a clipboard. If you detect low pressure, then high, then low again, that could be an infrasound wave!

A sequence of low and high pressures isn't necessarily a "wave"—it could also be a random pressure fluctuation in the air. That's why, to detect these sounds, researchers typically use an array of sensors spaced several meters apart. When an infrasound wave passes by a detector, it will pass over all the sensors at about the same time, which helps to separate out infrasound waves from random noise. If the sensors are spaced widely enough, you can even figure out which direction the sound came from by noticing which sensors registered it first.

Producing sounds like this would require a very large piano, because the strings would need to wobble back and forth so slowly that you could watch them move. (In a sense, a jump rope is just a stringed instrument with a frequency about five octaves below the lowest standard piano note.)

Even though we can't hear infrasound, it behaves like normal sound, carrying signals through the air. In fact, while ultrasound travels less far than normal sound, infrasound travels *farther*. An infrasonic signal with a frequency below 1 cycle per second—1 Hz—can travel all the way around the planet.

Sound recordings are sometimes plotted on a chart showing which sound frequencies were detected at which times. You can produce a plot like this from any sound recording, not just infrasound. In fact, the musician Aphex Twin has hidden "images" in his music that can be seen on a spectrogram.

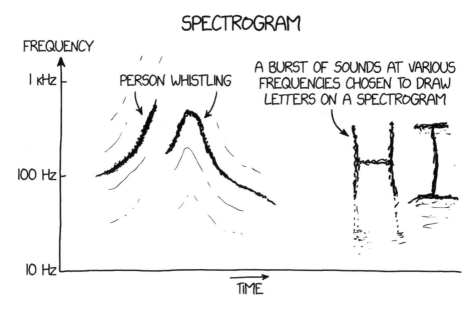

When a nuclear weapon goes off in the atmosphere, it creates a huge infrasound pulse. Much of the work on infrasound detection happened during the Cold War, as scientists built detectors to listen for those pulses. The last atmospheric nuclear detonation, as of this writing,[5] was a weapons test in China on October 16, 1980, so there haven't been any explosions for the networks to hear since then.

But an infrasound microphone picks up all kinds of interesting things beyond nuclear explosions. Large pieces of machinery that move rhythmically, like motors and wind turbines, create steady infrasound tones. Other infrasound notes are played by wind rushing over mountains, meteors entering the atmosphere, and even earthquakes and volcanic eruptions. A plot of atmospheric infrasound will also show warbling tones of unclear origin. It's just like regular sound frequencies — if you go somewhere quiet and listen very carefully, you'll hear all kinds of interesting noises, only some of which you can identify.

---

5 I *really* hope we don't have to revise this paragraph before the next printing.

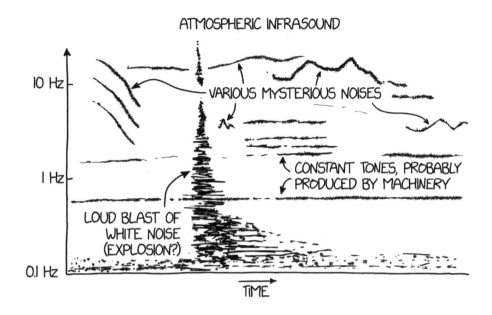

ATMOSPHERIC INFRASOUND

VARIOUS MYSTERIOUS NOISES

CONSTANT TONES, PROBABLY PRODUCED BY MACHINERY

LOUD BLAST OF WHITE NOISE (EXPLOSION?)

TIME

One of the most common infrasound tones is produced by waves on the open ocean. As the sea rises and falls, it presses rhythmically against the air, behaving like the surface of a huge, slow music speaker—the loudest, deepest subwoofer on the planet.

The sounds produced by the waves, called *microbaroms*, fall near 0.2 Hz. Playing microbarom frequencies on our piano would require an additional 75 keys, bringing the total to 235.

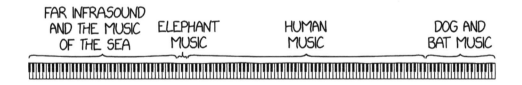

FAR INFRASOUND AND THE MUSIC OF THE SEA — ELEPHANT MUSIC — HUMAN MUSIC — DOG AND BAT MUSIC

That's a lot of keys. But if you master them all, you can play everything from Beethoven to bat hunting songs to the voice of the sea itself.

One last note: this piano will be difficult to construct. Piano strings won't work for producing ultrasound because the vibrations are too small and fade too quickly—even within the normal range of pitches, pianos typically need multiple strings for the highest notes in order to make them loud enough. Piano strings are also not ideal for producing infrasound: the strings would be too long to fit in a room and would have difficulty moving enough air around. For generating high and low notes, you'll want to use alternative techniques.

The most effective way to create ultrasound is through the *piezo-electric effect*, in which a crystal vibrates when you run electric current through it. The timekeeping element inside a digital watch or computer clock uses this effect: it contains a tiny piece of quartz shaped like a tuning fork that vibrates at a precise frequency in response to electric pulses. Similar quartz oscillators can be used to produce ultrasound of any desired frequency.

ROTARY
WOOFER
(INFRASOUND)

For the infrasound speaker, you might want to make use of a mechanism called a *rotary woofer*, a device that uses carefully controlled tilting fan blades to gently push air back and forth. By changing the pitch of the fan blades, it moves air forward, then backward, then forward again.

PIEZO
TRANSDUCER
(ULTRASOUND)

If you successfully manage to build the full 235-key piano, here's a sample piece for you to play. It will take some patience and it won't sound like much to your human ears.

But if any researchers out there are monitoring the atmosphere, listening for meteor explosions or nuclear weapons tests...

WHAT THE HECK?

... it will print out a stick figure on their spectrograph.

Infrasonata

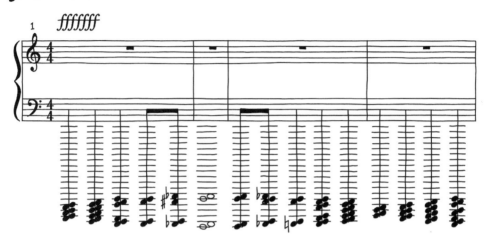

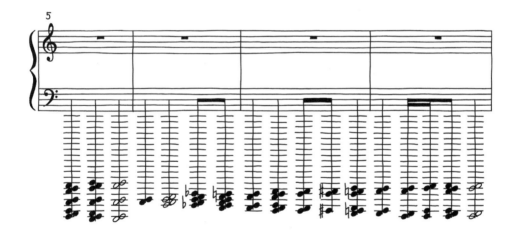

How to Listen to Music

IN MAY 2016, BRUCE SPRINGSTEEN PLAYED A CONCERT
IN BARCELONA. SEISMOLOGISTS AT THE NEARBY
INSTITUTE OF EARTH SCIENCES (ICTJA-CSIC) WERE
ABLE TO PICK UP LOW-FREQUENCY SIGNALS CREATED
BY THE AUDIENCE DANCING TO DIFFERENT SONGS.

ADAPTED FROM JORDI DÍAZ ET. AL., "URBAN SEISMOLOGY:
ON THE ORIGIN OF EARTH VIBRATIONS WITHIN A CITY", 2017

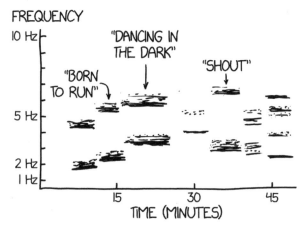

TOO BAD WE'RE STUCK IN THE
LAB TONIGHT. I WANTED TO GO
TO THE SPRINGSTEEN CONCERT.

CHAPTER 5

How to Make an Emergency Landing

A Q&A with test pilot and astronaut Chris Hadfield

How do you land a plane?

To answer this question, I decided to turn to an expert.

Colonel Chris Hadfield has flown fighter jets for the Royal Canadian Air Force and worked as a test pilot for the US Navy. He has flown more than 100 different aircraft. He also flew two Space Shuttle missions, piloted a *Soyuz*, became the first Canadian to walk in space, and served as commander of the International Space Station.

I contacted Col. Hadfield and asked if he could offer some advice on emergency landings, and he graciously agreed.

I made a list of unusual and improbable emergency landing scenarios, then called him up and posed each one to him to see how he would respond. I half expected him to hang up on me after the second or third question, but to my surprise, he answered every one with virtually no hesitation. (In retrospect, my plan to fluster an astronaut by throwing extreme situations at him might have been flawed.)

The scenarios and Col. Hadfield's answers—edited a bit for clarity and brevity, including some additional answers by email—are presented below. These aren't necessarily the only possible ways to handle each task, but they represent the first instinct of one of the world's foremost test pilots and astronauts, so they're probably a pretty good place to start.

COL. CHRIS HADFIELD

HOW TO LAND ON A FARM

Q: Suppose I need to make an emergency landing, and all I can see are farm fields. Which crop should I aim for? Should I pick a taller one that will provide more drag, like corn, or something low to the ground to give me a smoother surface? Would a field of pumpkins provide extra cushioning, like those barrels of water on a highway, or just make me more likely to flip over and catch fire?

A: I fly little airplanes, and that's something we think about all the time. When you're driving to your airfield, you look around, and think, how high are the beans? Have they brought in their hay? Has it rained recently? You can't land on a muddy field.

You want something where the crop is not so tall or thick that it will make your plane flip over. Obviously sunflowers would be a big mistake.

The best thing to land on is a freshly planted field. The worst place to land is somewhere that's just been plowed. Don't land on ginseng. They have to put up big sunshade structures and you'll get tangled up in them. You've gotta watch out for trees. Pastures are nice, but you have to be careful not to hit the cows. Corn is ok to land in up until the middle of June.

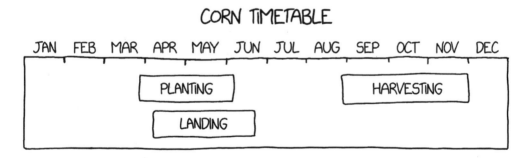

HOW TO LAND ON A SKI JUMP

Q: What if I'm making an emergency landing in a small plane, but the only open space I can find is an Olympic ski jump? What's the best way to approach it?

A: I was actually a ski instructor before I became a fighter pilot.

Olympic ski jumps are pretty dang high. There's that little flat section at the bottom, which would probably be your best bet. You could come in over the stands, nice and slow, and get down close to the ground, then just as the hill starts to rise in front of you, pull up. If you time it right, you could try to stall the airplane at the exact moment that you hit the slope. But you'd have to time it *just* right. If not, no do-overs.

HOW TO LAND ON AN AIRCRAFT CARRIER

Q: What should I do if I want to land on an aircraft carrier, but I'm in a passenger plane not designed for carrier landings? Should I try to snag my landing gear on the cable? How should I approach the carrier?

A: What you're going to want to do is get the captain of the aircraft carrier to turn her ship into the wind. Get the ship going as fast as she can make it go, which might give you 50 or 60 mph winds. For a lot of little airplanes, that's enough speed that it will get you going pretty slowly relative to the ship.

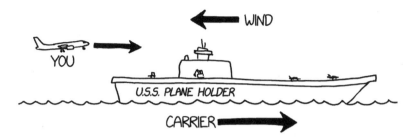

Get rid of those arresting cables; you don't want to snag one by accident. You need special gear to use an arresting cable. Unless you have a big strong hook, then you're going to want to do it completely aerodynamically.

Then you need to get yourself lined up. You're going to want to make use of every inch of the flight deck. You should extend the flaps, changing your wing from being flat to being sort of curved. If you watch birds land, they do that with their wings. When you try to fly slow, you deploy your flaps.

You want to hit the aircraft carrier RIGHT at the very back of the deck. Then you want to chop your power to zero, bring your engines back, and raise your flaps immediately. Otherwise wind can blow you off. However – *keep your hand on the throttle*. You want to be able to jam your throttle up and go around again. In fact, when military pilots land on aircraft carriers, they throttle up to full power right after they touch down, just in case the hook skips over the cable or the cable breaks.

One project I did was for the US Marine Corps. They were thinking, "What if we have an open space somewhere in the woods, but it's too short to land an airplane? Could we put a temporary arresting cable in the woods?" With a cable strung between big stakes, you can stop and land anywhere. I tested that system in Lakehurst, NJ.

HOW TO LAND ON A
HOSTILE AIRCRAFT CARRIER

Q: What if the captain doesn't *want* me to land? Would she turn to go downwind, to make it harder?

A: There's always stuff on the deck. If they don't want you to land, they can move stuff to block your way. There are lots of little carts that they use to tow the airplanes around, and they could just drive the little carts out all over the runway.

You'd have to sneak up on them and do it at the right time, and get lucky. You might be able to do it. But I don't think the captain would be happy. And now what? You've just landed in the most heavily fortified jail in the world, and declared yourself an inmate.

SO, UH... HOW'S EVERYONE DOING?

HOW TO LAND ON A TRAIN

Q: Could I land on a train by matching speed with it and gradually bringing the plane down to rest on the roof of one of the cars?

A: Yes, you can do that. Flatbed truck, too. You see that at air shows sometimes.

The hard part will be that, as you touch down, the train always moves up and down a bit, which will bounce you. That's the problem with landing on a truck, too. But it's absolutely doable.

HOW TO LAND ON A SUBMARINE

Q: Landing on an aircraft carrier sounds pretty easy. Could I land on a submarine?

A: Yes, if it's on the surface, driving fast into wind, and you have a slow, stable airplane. Like landing on a skinny, short, wet runway. I think it's been done. Sometimes hard to find a submarine when you need one, though.

HOW TO LAND FROM THE COCKPIT DOORWAY

Q: What do I do if I somehow accidentally close my sleeve in the cockpit door, and can't reach the front of the cockpit? But I can reach some objects—maybe trays of in-flight meals—that I can throw at the controls. If I'm good at throwing, could I land by hitting the right controls?

A: If it's a single-engine plane, no way. But in a plane with multiple engines, it might technically be possible. The way you're going to control things is power. If you have engines on each side, by moving the throttles up and down you can climb, and you can turn that way, too. If you're *really* careful about throwing the utensils, you can fly an airplane just by moving the throttles up and down.

There was a DC-10 that lost all hydraulics, over Sioux City, and the pilots managed to get control and steer that airplane all the way around to the runway using only the throttles.

HOW TO LAND A SPACE SHUTTLE IN DOWNTOWN LA

Q: In a scene in the 2003 movie *The Core*, Hilary Swank plays an astronaut in a Space Shuttle that has veered off course due to a navigation error. She realizes they're headed for downtown Los Angeles, and she plots a course to land in the Los Angeles river—basically a long flat-bottomed concrete canal. In the movie, they manage to land safely in the canal. Could something like that really happen?

A: The Shuttle touches down around 200 knots – 185 if you're light, 205 if you're heavy. You need a long straight runway, thousands and thousands and thousands of feet. We did the initial Shuttle landings on the huge salt flats of Rogers Dry Lake at Edwards Air Force Base. Once we got better at it, we started landing on a 15,000-foot runway.

What we really want to do is land where we take off, so we built a runway at Kennedy which was 15,000 feet long. The runway at Edwards is way out in the desert, so if you roll off the edge of the runway it's not so bad. The one at Kennedy has less room for error because it's surrounded by water and there are alligators.

When you're coming in to land at Edwards, you have to do the deorbit burn all the way back over Australia. The computer calculates the timing to bring you down over your landing site. But with enough planning, you could land on any long, straight, flat surface. Landing in Los Angeles drainage ditches? I'm not sure there are any that are long enough.

You might have to deorbit anywhere in the world. We identify every runway in the world. We carried a book in the Shuttle with diagrams of all of them. It's like a big picture book. It shows the orientation of the runway and everything.

HOW TO FIND A PLACE
TO LAND THE SHUTTLE

Q: If I'm not sure how to use the computer, could I just guess? Could I fire the engines somewhere over Australia, figuring that will bring me down in the right part of the world, and then look for a good landing site out the window as I get close? How much room in a landing do I have to improvise?

A: There's a fair bit of room! We fly big S-turns to bleed off energy. If we fly fewer turns we could fly farther. The closer you get, the less ability you have to change your mind. But it's not completely far-fetched. You have a chance, if you aim for a general area and eyeball things.

When they were flying the X-15, a Shuttle predecessor, pilots would try and make the test flights last as long as possible. Neil Armstrong ended up too low over Pasadena and had to land on the wrong lake bed. I'm glad he made it.

OOPS.

HOW TO LAND A PLANE FROM THE OUTSIDE

Q: Say I'm locked out, on the exterior of the plane, but I can crawl around and manipulate the flight control surfaces by hand.

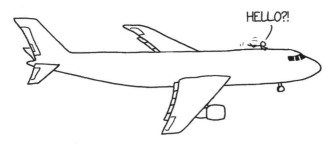

HELLO?!

A: People do wing walk, and occasionally people have done that to fix stuff. In an old, slow airplane, wind speed is low enough to stand on the wings. What you could do is use your own weight. You could control where the airplane is going by moving your weight around. Just move your weight to the right side, and the plane *might* begin a right turn.

If you can talk to the passengers inside, you could try to get them to run to the front or the back, and you might be able to control it a little bit that way.

But if you want to mechanically control the airplane, you have to get back to the tail. If you're on the wing you can only control roll—you can't control pitch or yaw. Roll is nice, but pitch and yaw are more important. To control pitch and yaw, you go back to the tail.

The problem is, you can't move those control surfaces by hand. No one is strong enough. If you were the Hulk, you might be able to find a handhold on the front of the tail with one hand and use the other to move the fin, and you could turn the plane left and right. Then you reach down and grab the elevator and do the same thing to control pitch. Conceivably, if you were good enough, you could use those to fly the airplane down.

Since you're not the Hulk, what you might do, if you were a little clever, is to find the *trim tab*. The trim tab is a little flat section on the edge of the surface that you use to make fine adjustments. You could move the trim tab, and it moves the whole elevator or the whole rudder.

HOW TO FLY THROUGH THE CHUNNEL

Q: Suppose I'm flying a very small aircraft like a Colomban Cri-Cri (wingspan: 16') over southern England right when Brexit happens. For complicated legal reasons, this means I need to land in France. Unfortunately, I'm a vampire who can't cross the water of the English Channel. Could I fly through the 25'-diameter Chunnel?

A: Yes – but 25' diameter and 16' wingspan means max clearance of 4.5' each side if you're right dead center. You could only climb or descend a few feet before your wingtips would hit concrete (you can do the math). The hardest part might be avoiding all the overhead wires at the Chunnel entrance and exit. And it would be dark, so you'd need to put lights on your Cri-Cri, or ask the nice Chunnel people to put all the lights on. But . . . for the tasty croissant and coffee at the landing aérodrome, it might be worth it.

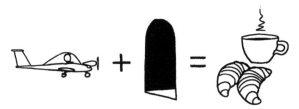

HOW TO LAND DANGLING
FROM A CONSTRUCTION CRANE

Q: If I'm flying an aircraft with a tailhook near a large construction crane, could I land by rolling onto my side and snagging the hook on the crane's dangling cable, then – once I've stopped swinging – have the crane operator gently lower me to the ground?

A: Maybe, if you're very lucky. Planes get snagged in power lines all the time and survive, where the crew have to be lowered by crane. But the inertia of your tailhook plane would likely be too much for the cable, and you'd snap it, plus even if you get snagged sideways, what keeps you from slithering down and hitting the ground? I'd go for power lines instead, and hope you don't cross the wrong wires and get electrocuted.

HOW TO GET OUT OF YOUR PLANE AND INTO ONE THAT HAS MORE FUEL

Q: Say my friend and I are flying a pair of small aircraft over a shark-filled ocean. I'm about to run out of fuel, but I have a parachute. My friend is flying alongside me. Could I get out of my plane and into theirs, then land *that* plane?

A: If they are open-cockpit biplanes, then maybe yes. You could trim your plane's controls to fly hands-off, get your friend very close, climb out on your wing, reach out and grab the wing of the other plane, and clamber into the cockpit. Needs to be an open cockpit so you don't have to deal with a canopy/doors, and a biplane so there are struts as handholds. If you jumped out of yours and hoped your buddy would somehow snag you as you floated under your parachute, my guess is you're shark lunch.

HOW TO LAND A SHUTTLE IF IT'S ATTACHED TO THE CARRIER AIRCRAFT

Q: Suppose I'm riding in the Space Shuttle while it's being carried by the Shuttle Carrier Aircraft. The carrier aircraft is on autopilot, but the pilot has decided to abruptly retire and has bailed out. What do I do? I assume if I have a parachute, I'd bail out from the Shuttle's exit hatch, but what if I don't? Should I try to detach the Shuttle, or get from the Shuttle into the carrier aircraft?

A: The initial flights of the Space Shuttle were drop tests from the Shuttle Carrier Aircraft. So I'd wait until you're within gliding range of a suitable runway, fire the separating mechanism from the SCA, pull back firmly to miss the SCA's tail, and then glide to a landing. Easy peasy.

THERE'S NO ONE AT THE CONTROLS
OF THE CARRIER AIRCRAFT, AND
THE SHUTTLE IS ON TOP WITH YOU
TRAPPED INSIDE. WHAT DO YOU—

FIRE THE SEPARATION
MECHANISM AND PULL
BACK TO AVOID THE TAIL.

SO WHEN DO WE GET TO
THE HARD QUESTIONS?

HOW TO LAND IN THE INTERNATIONAL SPACE STATION

Q: What should I do if I accidentally get left behind on the ISS when it's being deorbited? I know large objects occasionally survive uncontrolled reentry intact. If I find a parachute, where in the ISS should I hide to have the best chance of surviving to the point where I could parachute down?

A: You want a blunt, heavy piece of metal, and you need to have your own oxygen supply. So best bet is to get into a Russian Orlan spacesuit (you can easily self-don), get it running so you have pressure, cooling, and oxygen, jury-rig a parachute to it, and go into the FGB (Functional Cargo Block). Strap yourself to the thickest metal part near the middle, where there are the most massive things under the floor, batteries and structure, aligned with the solar array attachment points – and wait to see what happens. But … odds are slim to none.

Maybe bring your rosary beads so you have something optimistic to do while you wait.

HOW TO SELL PARTS
FROM A PLANE MID-FLIGHT

Q: Say I decide I want to land a plane, but first I want to sell as many parts as I can on Craigslist. I decide that shipping is too expensive, so I want to deliver them before I land by removing them from the plane and throwing them overboard as I pass over the buyer's house. How much of the plane can I sell and still land safely?

A: All the food. All the seats. But you have to be careful to keep your center of gravity in limits. If the center of balance is too far forward, then it becomes like a dart—no matter how hard you pull back on the stick, it's going to want to nose down. If the center of gravity gets too far aft, your aircraft becomes super unstable. Definitely get rid of all your cargo. Everything in the baggage compartment is something that someone paid to transport, so it's probably worth something.

HOW TO LAND A FALLING HOUSE

Q: When spacecraft like the *Soyuz* are returning to Earth, once they open up their parachutes they have no more control – you've described this phase as "falling like Dorothy's house." In *The Wizard of Oz*, when Dorothy woke up to find her house plunging toward Oz, can you think of anything she *could* have done to control her descent? Like if she looked out the window and saw the witch below, and wanted to miss her, or hit her, or aim for someone else?

A: I suppose she could have tried to run around and open windows and doors on different sides of the house, to see if she could have some aerodynamic control by changing the airflow. But I don't imagine it would be easy.

HOW TO LAND A DELIVERY DRONE

Q: Suppose I'm picked up by a malfunctioning quadcopter-style delivery drone, which hooked my jacket with its carrying arm and is heading up and out toward the ocean. I can extricate myself and climb up to reach the drone body, but how should I try to force it down gently without crashing?

A: Drones are battery-powered, so if I were you I'd pop the battery loose, let the drone fall a bit, stick the battery firmly back in, and play that until I can judge the descent, then choose a good moment to jump off. Best would be just after it got over water, in the shallows.

HOW TO LAND A ROC

Q: One last question. I know this might be outside your area of expertise, but suppose I'm picked up by a roc – the giant mythical bird. How should I try to force it to put me down without dropping me?

A: Your best bet is treat it like a big angry hang glider. If you pull yourself way off to one side, the roc will have to turn in that direction. If you somehow fling your weight forward, it will *have* to dive. If you were strong enough, you could sort of steer it, like a big, noncooperative glider.

The other thing you could do, if you have anything with you, like a tent or a lot of clothing, is you could deploy a parachute of sorts. Just the added drag of a parachute, or any big object hanging down, will irritate the heck out of any creature trying to fly. If you're a skydiver, deploy your parachute. You've always got your reserve chute.

You can start clipping its wings if you're armed. It depends whether you're willing to go on the offensive.

What you might want to do is be psychological. What does it want? Do you have food? What you don't want is for it to become irritated and let go of you. What it needs to do is be motivated to keep carrying you. I think I would try to get to a part of its body where it couldn't get me off. If I could get up behind its back and hold on, as long as you're holding on tight enough, it can't reach there. Like a bug it can't scratch. But if you're trying to modify its flight plan, either you've gotta use your own weight or use your own psychology or intellect. I don't know what motivates a roc.

Randall: Thank you so much for agreeing to answer these questions.
Col. Hadfield: Thank you for the … interesting … questions. I hope no one ever has to use my answers! But if you do – tell Randall so he can update this book.

CHAPTER 6

How to Cross a River

Humans like to live near rivers, which means we often find ourselves needing to cross them.

The simplest way across a river is to ford it—which effectively just means pretending it's not there, continuing to walk, and hoping for the best.

People usually try to ford rivers at areas where the river is pretty shallow, but even shallow water can be surprisingly dangerous. It's not always easy to tell how fast water is moving, and it only takes ankle-deep water to sweep a person off their feet.

If the river is too deep to ford, you can try swimming across. But whether swimming works or not depends a lot on the river conditions. If the river is moving too fast, you could be pushed around by the current, swept downstream, or sucked under obstacles or into rapids.

A normal person who knows how to swim – but isn't an athlete or anything – can probably move along at a few feet per second. This is much faster than some rivers and much slower than others – river speeds range from less than a foot per second to more than 30 feet per second.

If the river were an idealized region of water moving in a straight line at a constant speed, the time it took to swim across would be easy to figure out, since you could just swim directly toward the opposite bank and ignore the current. A faster-moving river would carry you farther downstream in the process, but you'd still reach the opposite bank in the same amount of time.

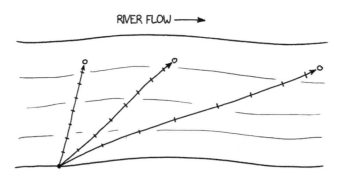

Unfortunately, real rivers don't flow at a uniform speed. Water tends to go faster in the middle than at the edges, and faster near the surface than at the bottom. The fastest-moving parts are usually over the deepest parts of the river, a little under the surface. For a smooth, uniform river flowing in a straight line, the speed might look like this:

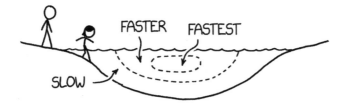

A riverbed with wide flat areas and deep channels might look more like this:

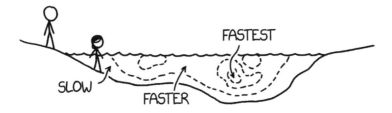

If you tried to swim across one of these rivers, your path would look more complicated. Moreover, real rivers don't flow in a straight line. They have eddies and whirlpools and currents that meander back and forth. In a real river, you might find the current keeps pushing you away from the bank, or sucks you under, or carries you downstream and over a waterfall.

That sounds dangerous. Let's take a look at some other options.

JUMP OVER IT

If swimming *through* the river doesn't appeal to you, you can try going *over* it. The simplest way, if the river is small enough, is to jump.

There's a simple formula for determining how far a projectile can fly when launched diagonally under ideal conditions.

$$\text{distance} = \frac{\text{speed}^2}{\text{acceleration of gravity}}$$

The exact distance you can jump depends on the details of your approach, launch, and landing, but this formula gives a pretty realistic estimate of what's possible. Based on the formula, if you run at 10 mph, you can expect to jump a gap of up to about 7 feet. This confirms that, for very small streams, jumping across is certainly an option.

You can increase your distance by increasing your speed, which is why champion long jumpers are sometimes also champion sprinters—in a sense, a long jumper is just a sprinter who's good at briefly going up instead of forward. Elite long jumpers can jump nearly 30 feet, which involves accelerating up to a sprinting speed of well over 20 mph shortly before takeoff.

Bicycles are faster than sprinters. If you get on a nice bicycle and pedal hard, you might be able to accelerate to about 30 mph. At this speed, you could—in theory—jump across a 60-foot river.

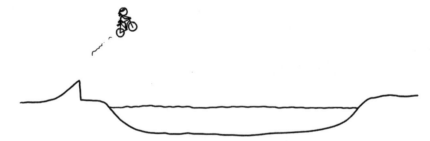

Unfortunately, thanks to conservation of energy, if you're going 30 mph when you take off, you'll be going 30 mph when you come down on the opposite bank. That's easily fast enough to cause serious or fatal injuries. It might actually be safer to try this stunt on a river that's *wider* than 60 feet. If you try to jump an 80-foot river instead of a 60-foot one, you'll land in the water near the far side, which would probably be less damaging to your body than landing on solid ground.

At least, assuming the water is deep enough.

NO DIVING

Faster vehicles can of course jump farther. A car going 60 mph could in theory jump a gap nearly 240 feet wide. However, landing a car at 60 mph is unlikely.

HELLO, FOLKS, THIS IS YOUR DRIVER SPEAKING. IS THERE ANYONE ON BOARD WHO KNOWS HOW TO LAND A CAR?

Motorcyclist daredevil Evel Knievel made a name for himself jumping over things with motorcycles, and famously attempted to jump the Snake River Canyon on a rocket bike which, for legal reasons, was technically classified as an airplane. Accounts differ on exactly how many bones Knievel broke during his career, but his ratio of SUCCESSFUL MOTORCYCLE JUMPS : BROKEN BONES was not large, and might have been less than one.

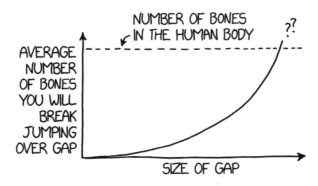

On second thought, maybe you should leave the jumping to the professionals, and then maybe the professionals should decide not to do it either.

GO ACROSS THE SURFACE

People can't walk on the surface of liquid water, at least not without the help of technology or supernatural forces.

There are viral internet videos of people running across water, riding bikes across water, and riding motorcycles across water. The basic principle behind all these stunts is simple: if you go fast enough, when you hit the water you'll ski across. These videos usually go viral because they seem at least sort of plausible, and the matter stays unsettled until the perpetrators of the hoax confess or the MythBusters try it out.

Here's a quick rundown of which types of stunts are real and which are fake:

WATER CROSSING METHODS
FROM VIRAL YOUTUBE VIDEOS

	FAKE	REALLY WORKS
RUNNING	✓	
BICYCLE	✓	
MOTORCYCLE		✓
SNOWMOBILE		✓

As people who do barefoot water skiing know, staying above the surface requires your feet to move at about 30 or 40 mph relative to the water. Even Usain Bolt's feet don't move that fast when he's sprinting.[1]

A bicycle won't work either. You can figure that out, without trying it, simply by asking an experienced cyclist. A cyclist can tell you that bicycles, unlike cars, generally don't hydroplane. They might slip on wet pavement, but because of the curved shape of the tires, which push water away to either side, a bicycle tire doesn't lose contact with the ground and "surf" on a layer of water.

---

1   If you were trying to stay above the water's surface by running, it would actually make more sense to run in place, so your feet would be moving quickly relative to the water's surface. A lightweight barefoot water skier with large feet can stay above the surface at only 30 mph, which is about 5 mph faster than the running speed of the fastest sprinters. So staying above water by running in place is probably impossible, but we won't know for sure until someone takes a champion sprinter—one with a small body and large feet—and lowers them slowly into a pool of water as they run in place. Good luck with *that* grant application!

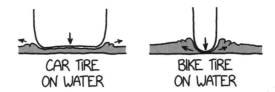

Motorcycles, which have flatter, treaded tires like cars', can hydroplane, and Myth-Busters has dramatically confirmed that they can also be used to cross short stretches of water. But that's taking us back into Evel Knievel territory.

Of course, there are specialized vehicles *designed* to travel over the surface of water.

If you have a boat, it's a perfectly good option. In fact, some rivers have boats permanently stationed to ferry people back and forth between opposite sides.

OTHER PHASES OF MATTER

When we said people can't run across water, that wasn't quite true. They can't run across *liquid* water. But water has other forms. Let's take a look at the other phases of matter, and see whether we could convert the river into them in order to make it easier to cross.

Freezing

To freeze a river, you'll need some refrigeration machinery and a source of power.

Thinking about the energy involved in freezing can be tricky. In a strict sense, turning water into ice doesn't *take* energy. When water freezes, it *emits* energy.

So if it takes energy to boil water, but freezing water gives off energy, why does our freezer use electricity instead of generating it?

The answer is that the heat in water doesn't want to leave. Heat energy naturally flows from warmer areas to cooler ones. When you put ice cubes in a warm drink, the heat leaves the drink and flows into the ice cubes, warming the ice while cooling the drink, bringing them both toward equilibrium. The second law of thermodynamics says that heat energy always wants to flow in this direction: the ice never spontaneously heats up the drink while getting colder. Moving heat from a *colder* area to a *warmer* one, against this natural flow, requires a heat pump, which takes energy to operate. When you try to remove heat from a river to lower its temperature and freeze it, you have to do work.

We can use estimates from commercial ice makers to figure out how much energy it would take to turn a river to ice through refrigeration. The US Office of Energy Efficiency and Renewable Energy, in its guide to estimating commercial ice machine power consumption, suggests a default estimate of 5.5 kilowatt hours (kWh) for every 100 pounds of ice produced. A normal spring flow rate for the Kansas River at Topeka might be 7,000 cubic feet per second, which gives us an estimated power of 87 gigawatts.

$$\frac{5.5 \text{ kWh}}{100 \text{ lb}} \times 1\frac{\text{kg}}{\text{L}} \times 7,000\,\frac{\text{ft}^3}{\text{s}} \approx 87\,\text{GW}$$

Eighty-seven gigawatts is a lot of power;[2] it's equivalent to the power output of a heavy-lift rocket as it takes off. Powering your refrigeration devices would take a similarly large generator, and that generator would take a lot of fuel. In fact, the flow rate of *fuel* into that generator would be about 300 cubic feet per second, which is nearly 5 percent of the flow rate of the river itself.

In other words, your freezing apparatus would need to be fed by a river of gasoline that is comparable in size to the river you want to freeze.

---

2   It's enough to go back to the future 71 times.

But maybe there's a way around this. Maybe you don't need to freeze the entire river. You could just freeze the surface.

As a general rule, ice needs to be at least 4 inches thick before it's safe to walk on. The Kansas river is about 1,000 feet wide, which will be the length of the bridge, so if we want to make the ice bridge 200 feet wide (to keep it from bending and breaking), then our ice bridge will be roughly 2,000 tons. Freezing that much ice would require about 330 megawatt hours of electricity, at a cost of about $50,000 (not counting the cost of all the ice machines.)

Boiling

We've covered the solid and liquid options. What about gas? Could you install some machinery upstream to convert the river from liquid to gas, then walk across the dry riverbed?

No, you couldn't. But let's find out why.

First, you'd need a way to heat the water. Obviously, you can't just use ordinary tea-kettles. Instead, you'd need to —

WAIT, WHY IS THAT OBVIOUS?

Ok. If you want to boil the Kansas river using ordinary teakettles, here's how to do it.

A typical teakettle holds 1.2 liters of water. Water has a high heat storage capacity—it takes a lot of energy to raise its temperature. But it takes a *huge* amount of energy to transform it from hot water to steam. Heating a liter of water from room temperature to 100°C takes about 335 kilojoules of energy. Pushing that 100°C liquid over the edge into becoming 100°C vapor takes a much larger 2,264 kilojoules.

You can see this effect when you boil water. It only takes about 4 minutes for most electric kettles[3] to heat water to a boil. But when you turn off the heat, most of the water is still there—it's at boiling temperature, but it's still in liquid form. If you want to fully boil the water—turn it completely to vapor—then you have to keep heating it for a total of about 30 minutes. This is much longer than the 4 minutes it takes to start it boiling.

The Kansas River's flow rate is 7,000 cubic feet per second, which works out to roughly 10 million teakettles per minute.[4] Since each kettle would need to boil its 1.2 liters of water for 30 minutes, you'd need a total of 300 million teakettles running in parallel to boil it.

If an electric kettle has a 7-inch circular base, you can pack them at a density of 3 per square foot.

Three hundred million kettles will take up a circular area 2 miles in diameter. To boil the river, you'll have to split it up and divert its flow across your kettle field. Each kettle will boil the water as it flows in, and once a kettle is empty, fresh water from the river will replace it.

Here's how this method would work in theory:

---

3   Most electric kettles—like most hair dryers—are limited to 1875 watts, because if they drew any higher, they wouldn't be able to be safely plugged into US household outlets rated at 15 amperes (amps, A).

4   Ten megakettles.

Now, here's how it would actually go:

Your electric kettles would draw roughly as much electricity as the entire rest of the country combined. There's no way you could get that much power focused on one location like this using our electric grid.

Which is probably for the best. Because if you could, things would not go well.

Boiling water creates hot steam. Steam rises. With a single kettle in a kitchen, that's fine—the steam rises, hits the ceiling, spreads out, and eventually disperses.

In a sense, this is also what would happen to your kettle field. But the experience would be a little more . . . dramatic. The column of steam would build into the stratosphere, spreading out and forming a mushroom cloud, like a volcanic eruption or nuclear explosion. When air rises, more air flows in from the side to take its place. You might not notice this when it happens over your stove with a single kettle, but the people living in Kansas around your kettle field would *definitely* notice. From all directions, winds would blow across the surface toward the kettles, converging on the base of the rising steam column.

Down at the base, things would not be going well. The kettles would be absorbing a huge amount of electrical energy and dumping it back out in the form of steam and heat radiation. The energy output of your kettle field would be greater than the heat output of a miles-wide lake of lava.

Heat is something of an equalizer. As a rule of thumb, anything that puts out as much energy as a lake of lava *becomes* a lake of lava. Your kettles would overheat, break down, and melt.

Say you manage to find fireproof, heat-proof kettles and wires. Then the kettles may start heating the lower layers of steam *too* fast. The heat would flow in faster than convection could carry it out, and the temperature of the steam would rise. Potentially, if you ran the kettle field long enough, the steam could begin turning from gas to plasma.

Here's what it will look like when you try to cross the river:

As you walk through the mud of the riverbed, you see a giant column of steam to your left, radiating intense heat, with a growing lake of lava at its base. From your right, a powerful wind blows along the riverbed. The wind cools you, for now, but if it gets too strong it might blow you toward the lava lake. From above, a gentle rain falls, turning the ground into warm mud. Overhead, electrical wires crackle and spark, as the entire US electricity grid shunts power into your lava lake.[5]

At this point, you realize: you never even needed to turn the kettles on. It took 30 minutes to fill them with water; you could have used that time to let a section of the river drain and walk across.

---

5 Once the kettles are removed, the crater they left behind will be filled in by the river, forming a temporary kettle-hole pond. (Thank you to the approximately four glacial hydrologists out there who laughed at that.)

But that wouldn't have been nearly as much fun.

KITES

If you don't have 300 million teakettles,[6] you may want to try crossing the river by kite.

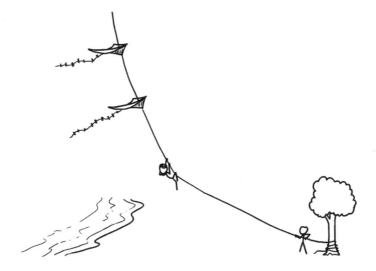

Kites actually have a bit of a history when it comes to river-crossing. When engineers wanted to build a suspension bridge across the gorge below Niagara Falls, they needed to start by getting a single cable to run from one cliff to the other.

---

6   For some reason.

They brainstormed ideas for how to get a cable across. They thought about having a ferry go across towing a cable, but the river was too turbulent and fast-moving for the boat to make it without being swept far downstream. The gap was too wide to shoot an arrow across, and they considered and rejected cannons and rockets. Ultimately, they decided to hold a kite-flying contest, and offered a $10 prize to whoever could fly a kite from one side of the gorge to the other.

After several days of effort, 15-year-old Homan Walsh succeeded in bridging the gorge. He launched his kite from the Canadian side and managed to get it snagged on a tree on the American side, winning the cash prize. The bridge engineers used the string to pull a stronger string across the gap, and after several more iterations, they had the two countries bound together by a half-inch cable.[7] Then, they started running more cables across the gap, built a pair of towers, and eventually constructed a suspension bridge.

Of course, if you're going to go the Homan Walsh route, you can simply cut out the middleman and fly *yourself* across on the kite. Human-lifting kites were briefly pioneered in the late 1800s and early 1900s, before the invention of the airplane made them seem slightly less exciting.

---

7   The July 13, 1848 issue of *The Buffalo Commercial Advertiser* contains the headline "Incidents at the Falls," under which appeared a breaking news story announcing that a very cute bird — a phoebe — has nested near the paddle wheel of the *Maid of the Mist* Falls tourist steamship, and has successfully raised and fledged a family of chicks several years in a row. I love old newspapers, and wish I got breaking news alerts on my phone about that kind of thing.

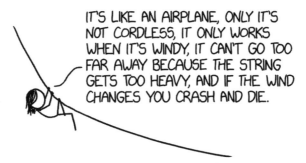

IT'S LIKE AN AIRPLANE, ONLY IT'S NOT CORDLESS, IT ONLY WORKS WHEN IT'S WINDY, IT CAN'T GO TOO FAR AWAY BECAUSE THE STRING GETS TOO HEAVY, AND IF THE WIND CHANGES YOU CRASH AND DIE.

Of course, not *every* flight on a human-lifting kite ends in a terrible crash due to changing winds. Sometimes, they crash for totally different reasons!

In 1912, Boston kite maker Samuel Perkins was testing a human-lifting kite in Los Angeles. He was soaring at a record altitude of 200 feet when a passing biplane sliced through the kite line. Miraculously, the fluttering kites acted as a parachute, and Perkins survived his plunge with minimal injuries.[8]

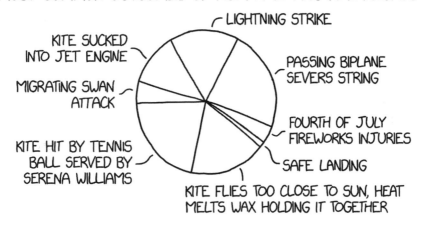

MOST COMMON OUTCOMES OF HUMAN-LIFTING KITE FLIGHTS

LIGHTNING STRIKE

KITE SUCKED INTO JET ENGINE

PASSING BIPLANE SEVERS STRING

MIGRATING SWAN ATTACK

FOURTH OF JULY FIREWORKS INJURIES

KITE HIT BY TENNIS BALL SERVED BY SERENA WILLIAMS

SAFE LANDING

KITE FLIES TOO CLOSE TO SUN, HEAT MELTS WAX HOLDING IT TOGETHER

You could also use a balloon in place of a kite. Balloons and kites are strangely similar—a balloon on a string is, in a sense, a kite mirrored around a diagonal line. A kite on a string "wants" to lie flat against the ground due to gravity, and wind flowing past creates an upward force on the kite. Its final diagonal angle is a compromise between the two forces.

---

8 The biplane wing was also damaged, but the pilot was able to land safely.

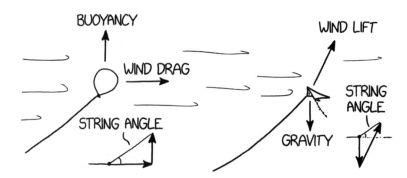

A balloon, on the other hand, "wants" to go straight up, and wind pulls it sideways. Again, the final angle is a compromise between the two forces. But as wind grows stronger, kites fly more vertically, while balloons fly more horizontally.

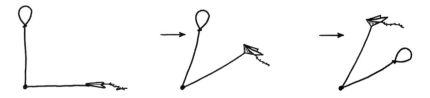

Once you're across the river, your challenge is getting down. But this is easy: for once, gravity is firmly on your side. You just have to make whatever's holding you up—kite, balloon, or other contraption—slightly less good at flying, and gravity will do the rest.

CHAPTER 7 → # How to Move

You've picked a place to move to, and now you need to get all your stuff there.

If you don't have very much stuff, and you're not moving very far, this will be easy. You can just put your stuff in a bag and carry it from your old house to your new house.

Unfortunately, if you have a lot of stuff, moving can be a lot of work. At some point in the move, many people take a look at all their possessions, realize how much work will be involved in moving, and realize that it would be easier to push everything into a hole and walk away, leaving it all behind. This is absolutely an option! Should you decide to take this route, turn to chapter 3: How to Dig a Hole.

If not, you'll need to pack up your stuff. The standard packing method, chosen by most people, is to put all your stuff in boxes and then carry the boxes out of the house.

Unless you're moving to your front yard, you're not done yet. You've only moved your stuff about 50 feet; depending on where you're moving, you could have hundreds of miles left to go. How do you get it there?

Carrying your stuff by hand is also a bad idea. Let's say you can walk while carrying about 40 pounds. As a general rule of thumb, all the furnishings and possessions in a typical 4-bedroom house will weigh around 10,000 pounds, which means you'll need to take a total of 250 trips.[1] If you have 3 people helping you, and you can walk 10 miles a day,[2] it will take you 7 years to move.

Things would be much easier if you could just make a single big trip with all the stuff at once. The good news is that, in a frictionless vacuum, pushing stuff sideways doesn't take any work at all. And if you're moving downhill, the move will actually require *negative* work – you'll get energy back! The bad news is that you probably don't live in a frictionless vacuum. Most people don't, despite the clear advantages one would offer when moving.

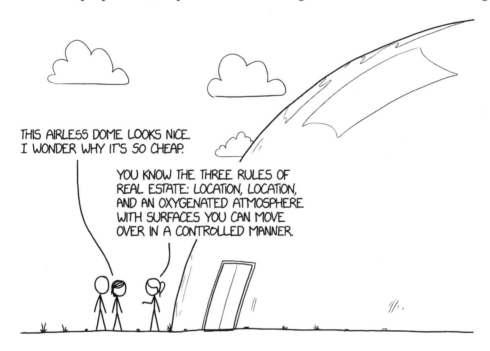

In our frictional air world, moving *does* take work. Your 10,000 pounds of stuff is heavy, and pushing it sideways takes force. The drag force exerted by the ground is simply the coefficient of friction between your boxes and the ground times the weight of the boxes. To estimate the coefficient of friction, we can see to what angle we have to tip it in order to make it slide, then calculate the inverse tangent of that angle.

---

1   You also may need to saw your refrigerator into pieces to make it light enough to carry.

2   On average. You'll probably be able to walk faster on the return trip, since you'll be walking without a load.

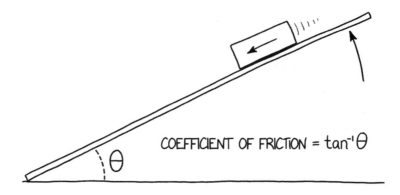

COEFFICIENT OF FRICTION = $\tan^{-1}\theta$

For a box sliding on a cement slab, the coefficient of friction might be 0.35, which means we'll need 3,500 lb of sideways force to move the boxes across the ground. This is too much for one person—it's about the force exerted by a 15-person elite tug-of-war team[3]—but it's within the capability of a large pickup truck.

OK, KEEP PUSHING!

Pushing a 10,000-pound load 200 miles takes about 5 gigajoules of energy, which is roughly equivalent to the electricity used by a typical house in 60 days. If you're using an elite tug-of-war team, that's 600 daily 2,000-calorie rations. Five gigajoules sounds like a lot, but it's not—it's only 40 gallons of gas.

Even if you have a truck powerful enough to push all your possessions across the country, this is probably a bad way to move. As the cardboard slides on the road, it will wear away, and your possessions will slowly be ground down.

---

[3]   Yes, there are elite tug-of-war teams. The sport is much more dangerous than is commonly realized; see *what-if.xkcd.com/127* for the horrifying details.

You can improve the situation by putting all the possessions on a sled made of some kind of hard, friction-resistant material. But you can improve the method even *more* by placing rollers under the sled that move with it. Now, add an axle, so you don't have to keep replacing the rollers. Congratulations, you've developed the wheel!

At this point, you've effectively reinvented the moving truck, which is the standard method for moving. But all that packing is still a lot of work. If you absolutely refuse to pack your stuff,[4] you have another option: move the entire house.

MOVE WITHOUT PACKING

Houses are relocated all the time. Sometimes, a house is moved to preserve it for historical reasons. Sometimes, it's cheaper to bring a vacant house from elsewhere than to build a new one from scratch. And sometimes, someone decides they want to move a house, and if they have enough money to do it, then they can just do it and they don't have to explain why.

Houses are heavy – a house weighs a lot more than all the stuff in it. A house's weight varies a lot, but could be around 200 lb/ft^2 including the foundation. Without the foundation, it might be significantly less. An average-size single-story home might weigh 150,000 lb, or 350,000 lb with the concrete foundation and/or concrete slab.

Lifting a house is difficult for reasons beyond the weight. A house might feel solid, but it can be less rigid than you expect. Some contractors compare it to lifting a king-size mattress – if you try to lift it from just one spot, only that spot will come up.

To lift up a house, you generally need to cut holes in the foundation and place I-beams under it, aligned with the load-bearing parts of the house. Then you can lift those beams, lifting the house with them.

---

4   Or to hire movers to pack the house for you.

You'll need to detach the house from its foundation first, which means removing any "hurricane ties" between the foundation and the frame of the house. Those ties are there to stop a hurricane from doing exactly what you're trying to do right now.[5]

Once you've lifted the house off its foundation, you'll need to find a vehicle to put it on—flatbed trucks are the most popular option. Then, you can use this truck to drive the house to its new location, assuming the roads are wide enough. Try not to take any turns too sharply.

_____

5   If there are no hurricane ties, you might be able to save yourself some effort—if you wait long enough, a hurricane or tornado might come along and move the house for you.

Driving a house is harder than driving a car.[6] Unless your house is unusually light-weight and aerodynamic, your gas mileage will probably suffer. How many miles per gallon can you expect? Well, we can estimate that using some basic physics. Modern internal combustion engines can convert about 30 percent of the input fuel energy into useful work. At highway speeds, most of the engine's work goes into fighting air resistance, so to figure out how much fuel your vehicle will consume, we just need to plug your house's parameters into the drag equation. (Since there are other sources of drag in addition to airflow, this is probably a best-case scenario.)

$$\text{gas mileage} = \frac{\text{gasoline energy density}}{\frac{1}{2} \times (\text{air density}) \times (\text{house cross-sectional area}) \times (\text{drag coefficient}) \times (\text{speed})^2}$$

$$= \frac{35\frac{\text{MJ}}{\text{L}} \times 30}{\frac{1}{2} \times 1.28\frac{\text{g}}{\text{L}} \times 18\text{ ft} \times 36\text{ ft} \times 2.1 \times (45\text{ mph})^2} = 0.8 \text{ miles per gallon}$$

I really love that we can ask physics ridiculous questions like, "What kind of gas mileage would my house get on the highway?" and physics has to answer us.

Drag increases quickly at higher speeds. If you're traveling at 45 mph, you'll get 0.8 miles per gallon. At 55, just a little faster, your efficiency will drop to 0.5 mpg. If you take your house out on the Autobahn and drive at 80 mph, you'll 0.2 mpg, burning a gallon of gas every 1,200 feet that you travel.

You probably shouldn't drive your house that fast, since 80 mph winds can remove important pieces of your roof. Even if you're under the speed limit, the police might not be happy about someone recklessly driving a house.[7]

---

6   For one thing, parallel parking is a huge pain. On the other hand, when you try to merge, people will probably be more likely to yield to you.

7   In order to transport a house on the highway, you generally need to obtain special "wide load" permits, and if you're moving your house based on instructions in a book by a cartoonist, it seems like a safe bet that you haven't applied for those.

If you *are* pulled over, you might try arguing that you're inside the house, and police can't come in without a warrant! In the United States, police officers are allowed to search vehicles based on probable cause, but not homes. It's the perfect crime!

The legal system may disagree. In the 1985 case *California v. Carney*, the Supreme Court ruled that motor homes and RVs, even when parked, qualified as vehicles and could be searched without a warrant. In their opinion, they cited *mobility* and *roadworthiness* as the key factors in determining whether something was a vehicle and could be searched.

> *The capacity to be "quickly moved" was clearly the basis of the holding in* Carroll, *and our cases have consistently recognized ready mobility as one of the principal bases of the automobile exception.*

—California v. Carney, 471 U.S. 386 (1985)

As far as I know, no court has ruled on the specific question of whether the vehicle exception applies to a house being transported on a flatbed truck, but be aware that you might be on shaky legal ground.

FLYING HOME

Maybe you've discovered some obstacles while planning your drive—like low overpasses or narrow roads. Maybe you don't want to apply for the "wide load" permits. Or perhaps you're in too much of a hurry to drive. If so, you could try flying.

Moving your entire house by air presents some challenges. The most powerful helicopters in the world can lift between 20,000 and 50,000 lb. That's enough to carry the 10,000 lb of possessions in a medium-size house, but not the house itself.

If one helicopter can't lift your house...could several? If you hitched multiple helicopters to your house and had them all lift at once, would they be able to lift a heavier load?

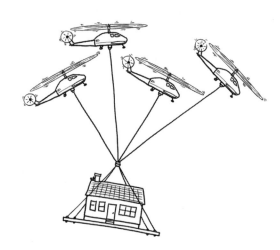

A multi-helicopter lift would present a few challenges. The helicopters would have to pull in different directions to avoid colliding, which would reduce their overall capacity. They would also need to coordinate carefully to avoid collisions. But you could solve both these problems by attaching the helicopters together rigidly, so they lifted as a single aircraft.

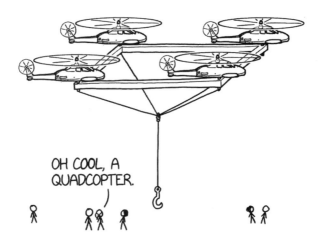

OH COOL, A
QUADCOPTER.

This idea sounds ridiculous, so, unsurprisingly, the US military studied it during the Cold War. In a 178-page report, they analyzed the idea of producing a super heavy-lift helicopter by the sophisticated engineering technique of taking two helicopters and gluing them together. The project[8] never went past the planning stages, possibly because the engineering diagrams looked a lot like mating dragonflies.

MULTI-HELICOPTER HEAVY LIFT SYSTEM
(1972 NAVAL FEASIBILITY STUDY)

MATING DRAGONFLIES

---

8   Which really should have gone by the code name HELICENTIPEDE.

Cargo airplanes can lift more than helicopters. A large aircraft like the C-5 Galaxy can lift nearly 300,000 lb, enough to carry a medium-size house–and possibly even the foundation, if the house is on the small side. *Size* might be more of a problem than *weight*–most houses are too large to fit inside a C-5 Galaxy's cargo hold.

There are a few whale-shaped specialty aircraft designed to carry unusually large pieces of cargo. The largest, such as the Boeing Dreamlifter and the Airbus Beluga XL, are built to transport pieces of other airplanes between factories while they're being constructed. If you ask nicely, maybe Airbus or Boeing will let you borrow one.

If you can't fit your house *in* an airplane, you could try putting it *on* one. That's how NASA transported the Space Shuttles across the country using a specialized Boeing 747 which carried the Shuttle on its back. To carry the Space Shuttle orbiter, the carrier aircraft has a special mount that protrudes from the top of the fuselage. This mount fits into a socket in the belly of the Shuttle orbiter. Next to the mount is an instructional plaque, which features the single best joke in the history of the aerospace industry:

ATTACH ORBITER HERE
NOTE: BLACK SIDE DOWN

Keep in mind that attaching your house to the outside of a carrier aircraft would subject it to 500 mph winds, which is far beyond what most structures are built to withstand. It would also probably affect the plane's handling.

There's another problem with moving your house by airplane: unlike a cargo helicopter, which can take off and land vertically, an airplane can't move your house without knocking down a lot of telephone poles, trees, and neighboring houses. If you don't live at the end of a runway, taking off will be a problem.[9]

But if all you want to do is push your house into the air and then push it sideways, why do you need the entire airplane? Why not just the pushing part? The engines on a 787 Dreamliner can produce 70,000 lb of thrust and only weigh 13,000 lb, which means two of them could lift a small house into the air. The application of this is obvious.

---

9   If you *do* live at the end of the runway, I would love to know about your homeowner's insurance policy, and whether your comprehensive auto insurance covers collisions with aircraft.

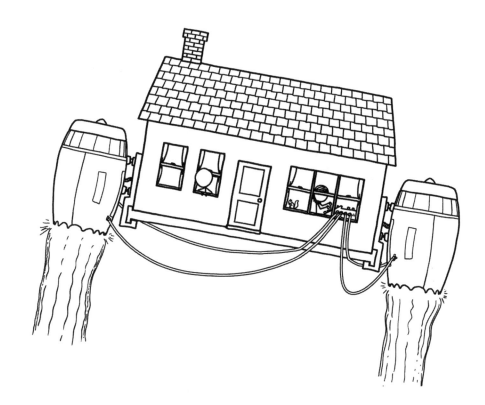

You might think that airliner engines wouldn't be very good at hovering in place. After all, engines need oxygen to burn, which they pull in through those big intakes in the front. It seems like they should be less efficient at scooping up air when they can't use their forward motion to help. But most turbofan engines produce their maximum thrust when they're sitting still. At higher speeds, the engine does take in air more efficiently, but the extra drag from all that incoming air counteracts the extra thrust the engine produces. Only at very high speeds, near Mach 1, does the ram effect cause the engine thrust to increase again.

Two engines might, in theory, be enough to get your house into the air, but you probably want to add a third and fourth for safety and stability.

Ok, you've got your house in the air. How long can you hover and fly around like this?

When hovering, jet engines need a lot of gas. At full power near sea level, each one consumes almost a gallon of jet fuel per second. Carrying more fuel lets you hover longer, but it also means you're heavier. If you add too much fuel, you'll be too heavy to lift off.

To figure out how long this kind of vehicle can hover if loaded with the maximum amount of fuel, you multiply the engine's specific impulse by the natural log of its thrust-to-weight ratio. This gives you the amount of time the engine can hover when it starts with a full load of fuel.

$$\text{hover time} = \frac{\text{engine thrust}}{\text{mass flow rate} \times \text{gravity}} \times \ln\left(\frac{\text{engine thrust}}{\text{engine weight}}\right)$$

For a large modern turbofan engine hovering in place at sea level, this number comes out to a little over 90 minutes. When you add on the extra weight of your house, it means your flying time will be *less* than 90 minutes, no matter how many engines you add. If you limit your horizontal speed to 60 or 70 mph, and you're moving farther than 100 miles, then you'll need to stop for fuel along the way.[10]

---

10 If you do run out of fuel mid-flight and start to fall, please consult chapter 5: How to Make an Emergency Landing, and skip to the section "How to Land a Falling House."

MOVING IN

When you arrive at your new home – or your old house arrives at its new location – there's a huge amount of work left to do. If you've brought your entire house, you may have to dig a foundation,[11] and if there's an existing foundation, you'll need to connect your house to it and firmly affix it. If there's already a house on the foundation you want to use, make sure to remove it before putting your own house there. Just send someone ahead of you with another set of engines and have them repeat the steps above on the house at your destination. Once you reach the point where the house is in the air, turn up the jet engines to full and jump out. After that, you can stop worrying about it; it's somebody else's problem now.

11 See chapter 3: How to Dig a Hole.

After moving in, you may need to set up the utilities, like heat, water, and power.[12] If you're feeling particularly civic-minded or excited about becoming part of your new community, you may want to introduce yourself to your new neighbors.

UNPACKING

If you've brought your possessions in moving boxes — or packed them up to secure them during flight — then you'll have a lot of work to do. You'll need to set up your furniture so you have a place to put your stuff, then unpack the boxes full of stuff and figure out where everything goes, which may involve a lot of trial and error.

---

12   See chapter 16: How to Power Your House (on Earth).

If unpacking seems too daunting, you can adopt a strategy that has probably been popular as long as humans have been moving their homes from one place to another: clear enough room to put your mattress on the floor, unpack the one box that has your toothbrush and phone charger, and worry about the rest in the morning.

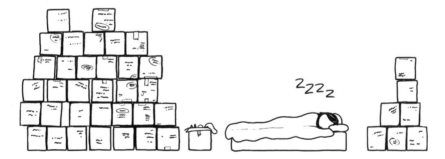

CHAPTER 8

How to Keep Your House from Moving

Once you're settled in to your house, you generally want it to stay where it is.

If you're worried that the house will blow away, or that some prankster will attach jet engines and send it blasting off into the distance, you can attach it to its foundation with hurricane ties. The foundation can also be anchored to the bedrock below using long metal piles.

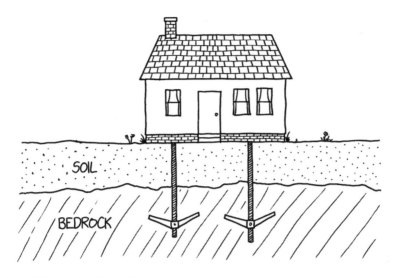

But what if the bedrock itself moves?

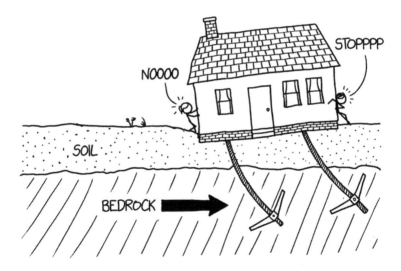

Tectonic plates are in constant motion. Much of North America is moving west, relative to the rest of the Earth, at about an inch per year. It seems obvious that property lines must move with the crust, since the alternative would be ridiculous—an inch of plate movement per year is enough that within just a decade or two, you could lose your garden on one side of your house while taking ownership of your neighbor's on the other.

Rather than being defined by coordinates, geographic boundaries are generally anchored to the ground. As a general rule, the final legal authority on the exact location of a boundary is typically *not* a set of coordinates or the text of the agreement creating the border – it's the markers left by the original survey based on that agreement, along with documentation created by the surveyors that can be used to reconstruct the location of each marker if it's moved or destroyed.

The International Boundary Commission, the body in charge of managing the US-Canada border, periodically publishes updated coordinates for the border, but their publications don't change where the border *is* – they just provide everyone with better information about it. The actual border is defined by "boundary monuments" – usually granite obelisks and steel pipes driven into the ground – along with photos and surveying information. If the land moves, the borders move with it, and the coordinates need to be updated.

To reduce the need for these updates, different countries and organizations often use slightly different latitude and longitude grids – geodetic datums – which are anchored to a particular tectonic plate. These grids move with the plate, and may differ from one another by several meters or more. Thanks to these different grids, no latitude/longitude coordinates are ever really precise and unambiguous without lots of information about the datum they're in. If you think that sounds like a huge headache for anyone who has to deal with precise coordinates, you're right.

By using a continent-specific grid, governments and property owners can partly mitigate the problem of the ground drifting out from under the grid – but that doesn't solve it completely, because sometimes one part of a continent is moving relative to another.

If your house is located on a plate boundary like the San Andreas fault, one part of your yard might be moving past the other at over an inch per year, and boundary monuments might start to disagree. Does your yard gradually split into two pieces? Could your house drift off your lot completely?

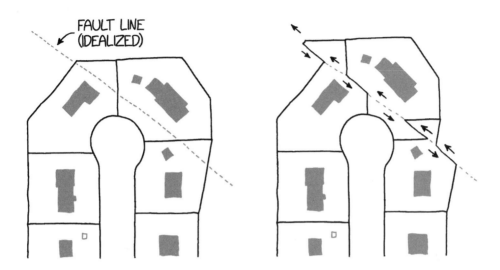

The 1964 Alaska earthquake shifted much of the city of Anchorage sideways by roughly 15 feet. To deal with the resulting land-ownership questions, the state passed a law in 1966 allowing all property lines to be resurveyed to match the new location of the ground. California passed a similar law, the Cullen Earthquake Act, in 1972, which let property owners ask courts to redraw lines in a way that protected the interests of all those involved.

If you live in Alaska or California, at least, it might seem like these rules would keep your neighbor from gradually taking ownership of parts of your house. But there's a catch: courts have ruled that these laws apply only to *sudden* movements, not gradual ones.

In the 1950s, road construction in the coastal town of Rancho Palos Verdes, California, caused an entire neighborhood to begin gradually migrating downhill in a creeping slow-motion landslide. By the end of the 20th century, the neighborhood had moved several hundred feet, putting some houses on property claimed by the city. The city told the homeowners to leave, but some occupants, including homeowner Andrea Joannou, took the city to court to ask that the property lines be redrawn. In 2013, in the case *Joannou v. City of Rancho Palos Verdes*, the courts ruled in favor of the city, determining that since the land movement was not caused by a sudden and unforeseen event, the homeowners could have taken steps to respond—presumably, steps such as anchoring the homes to bedrock, or having them moved back up the hill every few years.

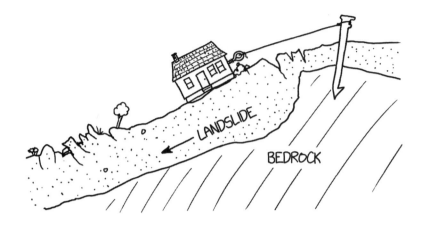

If the bedrock itself is moving, you could find yourself in unsettled legal territory. If there are established boundary monuments nearby, and they moved along with you, then you can argue that your property is anchored to those monuments. After all, they're the final authority on property lines. But if the boundary monuments are far away or have gone missing—as is often the case—then your property may be defined only by coordinates relative to some larger grid, and you might conceivably find that your plot of land has drifted into someone else's possession.

In that case, your best approach might be to try to buy the land on the *far* side of the neighbor's house. That way, if your neighbor takes possession of part of your house, you can simultaneously take possession of part of theirs.

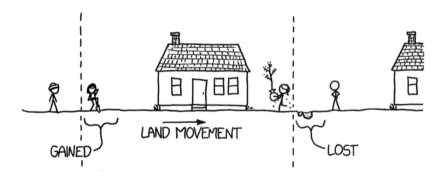

But when it comes to schemes to apply property line rules in unusual circumstances, you may want to be cautious. In their 1991 decision in *Theriault v. Murray*, Maine's Supreme Court said that boundaries are determined "... in descending priority, by monuments, courses, distances, and quantity, **unless this priority produces absurd results**" [emphasis added].

If you do end up in court with your neighbor...

... the judge may decide the situation qualifies.

How to Chase a Tornado

(without leaving your couch)

I WANT TO GET INTO STORM CHASING, BUT I'M NOT IN ANY RUSH AND THIS COUCH IS REALLY COMFORTABLE.

IF YOU SIT AND WAIT LONG ENOUGH, EVENTUALLY A TORNADO WILL COME TO *YOU*. THIS MAP SHOWS HOW LONG YOU WILL HAVE TO SIT—ON AVERAGE—BEFORE AN EF-2 OR STRONGER TORNADO PASSES DIRECTLY OVER YOU.

(ADAPTED FROM CATHRYN MEYER ET. AL., "A HAZARD MODEL FOR TORNADO OCCURRENCE IN THE UNITED STATES," 2002)

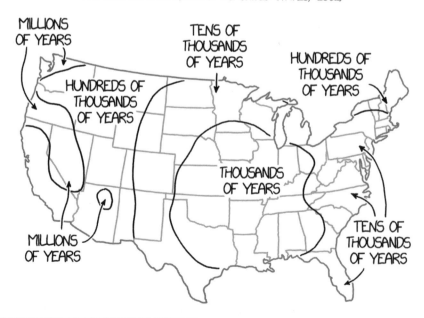

CHAPTER 9

How to Build a Lava Moat

There are lots of reasons for wanting a lava moat around your house, some more practical than others. Perhaps you want to deter burglars, keep ants from getting inside, or stop neighborhood kids from stealing pies that you leave cooling on your windowsill. Or maybe you want to give your landscaping more of that "medieval supervillain" aesthetic and create some excitement for your neighbors, fire department, and local zoning board.

MAKING LAVA

It's actually pretty easy to make lava, at least in principle – the ingredients are just rocks and heat.

LAVA
NUTRITION FACTS

SERVING SIZE: 1 KG
SERVINGS PER VOLCANO: VARIES

TOTAL CALORIES: 350 (HEAT)

	% DAILY VALUE*
TOTAL FAT: 0g	0%
SATURATED FAT: 0g	0%
TRANS FAT: 0g	0%
CHOLESTEROL: 0g	0%
SODIUM: 28g	1,200%
TOTAL CARBOHYDRATES: 0g	0%
DIETARY FIBER: 0g	0%
SUGAR: 0g	0%
PROTEIN: 0g	0%

CALCIUM: 3,500%	IRON: 250,000%
MAGNESIUM: 5,000%	ZINC: 450%

*PERCENT DAILY VALUES ARE BASED ON A NORMAL DIET WHERE YOU DON'T EAT LAVA

Most rocks melt at temperatures between 800°C and 1,200°C. That's hotter than a household oven, but achievable using a high-temperature furnace, charcoal forge, or even a giant magnifying glass.

To provide the actual material for your lava, you could try using whatever rocks you find lying around, but be careful; some rocks will melt or explode when heated due to trapped gases. The Syracuse University Lava Project, which produces artificial lava for both geologic research and art projects, uses billion-year-old basalt from Wisconsin. The basalt formed when the core of the North American continent developed a crack down the middle and large amounts of magma bubbled through. The crack eventually healed up, but it left a crescent-shaped scar of dense basalt buried beneath the soil of the Midwest.

If all you care about is having a moat that sets things on fire, you don't necessarily need to stick with volcanic rocks at all; you could also try using the sorts of molten glass used in glassblowing, or a metal with a reasonable melting point, such as copper. Aluminum's low melting point makes it an appealing moat material, but it melts at a low enough temperature that it doesn't really glow – and it's not really a lava moat if it's not glowing ominously.

KEEPING LAVA MOLTEN

Keeping lava molten is difficult because lava constantly radiates energy in the form of light and infrared radiation. Without a steady supply of heat, the lava will quickly cool down and solidify. This means you can't simply melt your lava, pour it into your moat, and call it a day. To keep it from cooling down and solidifying, you'll need to supply a steady flow of heat energy to make up for the losses.

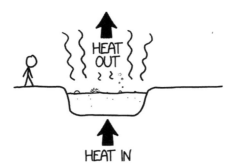

Your moat is going to need some kind of built-in heating apparatus.

You can think of a lava moat as a long, skinny, high-temperature open-top furnace. Industrial furnaces of this sort are often gas heated, but there are also electric versions that use high-temperature heating coils. Gas heating may cost significantly less, but electric furnaces tend to be simpler and offer more-precise temperature control. Regardless of the power source, the basic design is the same: a crucible to hold the lava, a heating coil or hot gas jet to heat the crucible, and insulation around that.

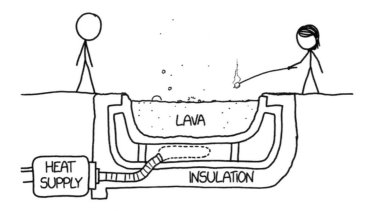

How hot does our lava need to be? We can pick materials with a low melting point to reduce energy consumption, but if the temperature is too low, the moat won't glow.

For something to glow from heat, its temperature has to be above about 600°C, and if you want a really nice bright orange-yellow color visible in the daytime, like the kind of lava you see in the movies, you'll need a temperature above 1,000°C.

We can use research on real lava flows to estimate how much energy a moat will radiate when the lava reaches a particular temperature, which tells us how much energy you'll need to supply to keep it molten.

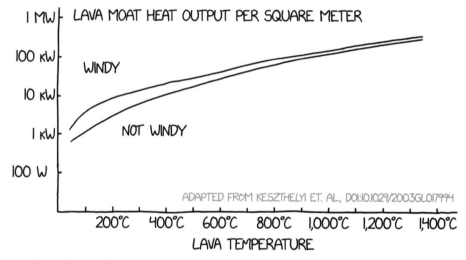

The chart above tells us that a 900°C lava pool will radiate roughly 100 kilowatts of heat per square meter. If electricity costs around $0.10 per kilowatt-hour, then each square meter of a 900°C lava moat will cost at least $10 per hour if heated electrically. If your moat is a meter wide and encloses an area of one acre, it will cost roughly $60,000 per day to keep it molten.

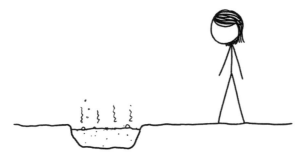

A 1-meter-wide moat might seem too narrow to deter human intruders,[1] since people can normally hop over a gap of that size without too much trouble. However, the heat of the lava moat will be dangerous even if they don't fall in. Near the surface of the lava, the heat will be intense enough to cause second-degree burns in less than a second. Even approaching the lava may be difficult. For someone standing a few meters away, the heat flux would be pretty high—enough to cause pain in exposed skin within 10 seconds, according to firefighter safety guidelines.

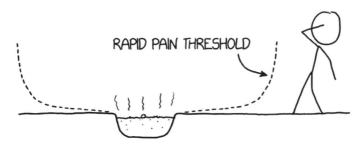

RAPID PAIN THRESHOLD

A 1-meter moat isn't impenetrable; someone wearing thick clothes and boots could conceivably jump over the moat without injury, as long as they avoided falling in and didn't linger too long on either side.

You could deter moat-jumpers by making the moat wider or the lava hotter. Both of those options will increase the expense, as shown by this approximate price table:

---

1 Your lava moat might help with keeping ants out, but it could also attract lava crickets. These insects, *Caconemobius fori*, live on or near recently cooled lava flows. Not much is known about them, since they are—as you might imagine—challenging animals to study.

LAVA MOAT HEATING PRICE GUIDE
(ENCLOSED AREA: 1 ACRE)

WIDTH	TEMPERATURE		
	600°C	900°C	1,200°C
1m	$20,000	$60,000	$150,000
2m	$40,000	$120,000	$300,000
5m	$100,000	$300,000	$750,000
10m	$200,000	$600,000	$1,500,000

COOLING

So far, we've only discussed the cost of heating the lava. But if you're going to live in the middle of this lava moat, you'll need to worry about cooling the house as well. Even with a sizable gap between the moat and the house, the lava's heat radiation will eventually make things uncomfortably hot in the house. If the side of the house is 10 meters from the moat, and you stand near a window, the heat radiation will exceed firefighter thermal exposure limits.

You can reduce the amount of thermal radiation reaching the house by recessing the moat into the ground, so more of the heat radiates upward. But this will only partially solve the problem; the ground around the moat will still be pretty hot, and radiate heat toward you. If there's a breeze, it will carry a stream of hot air downwind from the lava—and that's an inherent problem with lava moats: no matter which way the wind is blowing from, you're *always* downwind.

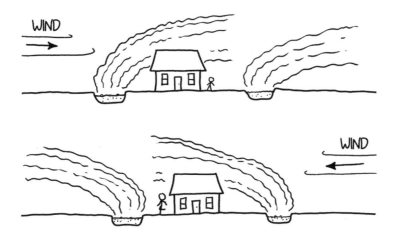

Fortunately, cooling your house is easier than heating your moat. If you have a source of cold water, like a nearby spring or river, you could run the water through your walls to carry away excess heat. Water's huge heat-storage capacity means that it can be used to remove a lot of heat with only minor pumping costs. This strategy has been used by tech companies to cool server rooms: Google, for example, has a data center on the coast of Finland cooled by seawater.

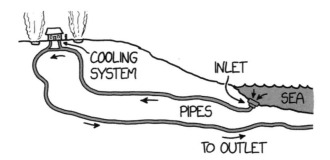

You may also want a source of ventilation from outside the moat, especially if your particular lava mix is prone to giving off toxic gas. Luckily, the heat of the lava can help you out here—if you install ventilation tunnels under the moat, the rising air from the lava will tend to suck air in through the low-level tunnels. This "natural draft" effect is used in industrial cooling towers like those over nuclear reactors and may reduce the need for fans to bring in cold air.

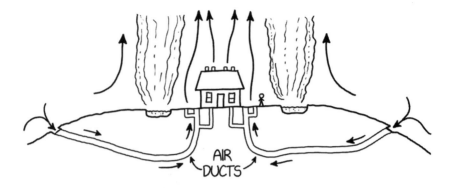

But be careful—if your water-cooling system takes in water from the ocean, you may find it unexpectedly clogged; nuclear reactors have sometimes gone into emergency shut-down when their intakes are blocked by swarms of jellyfish.

The jellyfish may point to a deeper problem with lava moats. Installing a lava moat provides additional protection, but the moat requires additional infrastructure, which creates its own vulnerabilities.

The jellyfish-clogged water intakes are bad enough, but from a supervillain point of view, perhaps you should be even more worried about the network of air ducts beneath your house. Because if there's one thing we've all learned from action movies...

... it's that someone always ends up sneaking through the ventilation shafts.

CHAPTER 10

How to Throw Things

According to a well-known legend, George Washington threw a silver dollar over a big river.

Like many anecdotes about Washington, this claim didn't spread until after his death, and the details are hard to pin down. Sometimes it's a silver dollar, sometimes it's a rock. Sometimes the river is the Rappahannock, sometimes it's the much wider Potomac. All we can say for sure is that people really liked telling stories about Washington, and apparently "throwing something over a river for no reason" was seen as a heroic deed.

It's not clear why throwing a silver dollar over a river is a good qualification to be president, but people seemed impressed by it. It's a shame that the story didn't become popular until after his death, because the campaign ads would have been pretty good.

Which objects could Washington have thrown over which rivers? How would he stack up against other presidents, and other non-presidents?

Let's look at a very abstract representation of what happens when a person throws something:

1. Person has the object
2. ???
3. The object is flying away

Oddly, even without knowing what happens in step 2, it turns out we can come up with a pretty good guess about how far someone can throw an object by looking at the constraints placed on them by physics.

Human bodies are limited in size. Whatever the thrower does to the projectile has to happen within a small region of space around their body.

PLAYER IN HERE SOMEWHERE

In order to throw something, a person needs to accelerate it using their muscles, and human bodies can only exert so much muscle power at once. Across a variety of sports, from rowing to cycling to sprinting, the power output that top athletes can deliver to an object over a short burst—such as a rowing stroke—is usually limited to around 20 watts per kilogram of body weight. This suggests an athlete who weighs 60 kilograms might have 1,200 watts of power available for throwing.

Let's assume the athlete "throws with their whole body," delivering that full power output to the ball over the short distance before it leaves their grasp:

WORK DONE HERE

Under these assumptions, we can use the equations of motion under constant power[1] to determine the ball's final speed:

$$\text{speed} = \sqrt[3]{\frac{3 \times \text{throw motion length} \times \text{body mass} \times \text{power}}{\text{ball mass}}}$$

If we plug in the average weight of an MLB baseball pitcher (208 lb) and the mass of a baseball (5⅛ oz), and we assume the throw motion covers a span similar to their height (6 feet 2 inches), we should be able to figure out a very rough estimate for a pitcher's fastball speed:

---

1   Elementary physics classes commonly analyze motion under constant *force*, and students may see those equations so often they learn them by heart. The equations of motion under constant *power*, with different exponents and coefficients, are a little more obscure. They're outlined in a 1930 paper by Lloyd W. Taylor of Oberlin College titled *The Laws of Motion Under Constant Power*.

$$\text{speed} = \sqrt[3]{\frac{3 \times 6'2'' \times 208 \text{ lb} \times 20 \frac{W}{kg}}{5\frac{1}{8} \text{ oz}}} = 94 \text{ mph}$$

Ninety-four mph is almost *exactly* the average speed for a four-seam fastball! Not bad for a formula that knows nothing about the pitcher.

If we put in numbers for a quarterback and a football, we get 67 mph. That's a little faster than actual football passes, which top out around 60 mph, but it's not far off.

$$\text{speed} = \sqrt[3]{\frac{3 \times 6'3'' \times 225 \text{ lb} \times 20 \frac{W}{kg}}{15 \text{ oz}}} = 67 \text{ mph}$$

Sadly, the precision of our answer is probably just coincidence, since this model has a problem.

According to our equation, extremely light balls can be thrown arbitrarily fast: a baseball that weighed half an ounce could be thrown at 200 mph! In reality, a baseball pitcher can't deliver all their power to the ball—they need to accelerate parts of their hand and arm to high speeds along with it.

To account for the hand speed limit, we can add a small "fudge factor," tweaking the formula by adding a little bit of weight to the ball—equal to 1/1000th of the player's body weight—to represent the weight of the fastest part of their hand. This places an upper limit on throwing speed for lightweight objects, which is consistent with reality, without distorting the results for heavier objects too much.[2]

We can combine this with an approximate equation for the distance a projectile will fly through the air[3] to produce a **unified theory of people throwing stuff really far:**

$$v = \sqrt[3]{\frac{3 \times \text{thrower height} \times \text{thrower weight} \times \text{power output}}{\text{ball mass} + \frac{\text{body mass}}{1,000}}}$$

Power output: 20 W/kg for a trained athlete, 10 W/kg for a normal human

$$v_t = \sqrt{\frac{2 \times \text{ball mass} \times \text{gravity}}{\text{cross-sectional area} \times \text{air density} \times \text{drag coefficient}}}$$

2 Note: The formula now underestimates a baseball player's pitching speed as being only about 80 mph, but it gives otherwise reasonable results. The discrepancy might be explained by the fact that baseball players spring forward, giving them some forward speed at the start and stretching their pitch over a longer distance, but it's a very simple model—we don't want to go too far trying to explain or correct every deviation.

3 This equation is based on approximations from the 2017 paper *Approximate Analytical Investigation of Projectile Motion in a Medium with Quadratic Drag Force* by Peter Chudinov. If the projectile is dense or the atmosphere is thin, it's equivalent to the standard range equation for an object launched at a 45° angle (range = v^2 / g), but at higher speeds, where air resistance is a bigger factor, it gives shorter distances.

$$\text{range} \approx \frac{v^2\sqrt{2}}{\text{gravity}\sqrt{\frac{4}{5}\frac{v^4}{v_t^4} + 3\frac{v^2}{v_t^2} + 2}}$$

v = throw speed, v_t = terminal velocity

This model isn't perfect. It's an unwieldy set of equations, and it's based on just a few input variables and extremely simple assumptions, so it can't be more than an approximation. We could make it much more accurate by putting in a more specific model of throwing mechanics, or more accurate data about pitchers. But if we make the model more specific, it becomes narrower in what we can apply it to. What's really fun is how broad it is. We can plug in *anything*.

Sure, we can use it to figure out how far a quarterback can throw a football. The longest NFL passes tend to travel 60-something yards through the air, and our equation gives a result that's pretty close.

(NFL quarterback, football) → 73 yards

But we can also use it to figure out how far a quarterback can throw *other* objects. Let's try an 11½-lb Vitamix 750 blender:

(NFL quarterback, 11½-lb blender) → 18 yards

All we need is a rough idea of the blender's weight, shape, and drag coefficients.

We're not limited to quarterbacks, either. We can plug in anyone whose height and weight we can estimate.

(Former president Barack Obama, olympic javelin) → 97 feet

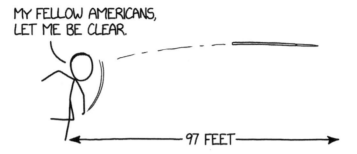

(Singer Carly Rae Jepsen, microwave oven) → 12 feet

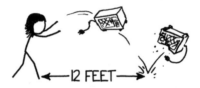

You can play with this calculator at *xkcd.com/throw*.

Using this formula, along with your height, weight, and level of athleticism, you can figure out how far you can throw arbitrary objects.

WASHINGTON'S THROW

What does our model say about George Washington's silver dollar feat?

Washington was famously athletic and liked throwing things—he reportedly threw a rock to the top of Virginia's Natural Bridge from the river below—so we'll give him a power ratio of 15 W/kg. That puts him halfway between a normal person and a trained elite athlete.

NORMAL PERSON   GEORGE WASHINGTON   ELITE PRO ATHLETE

The drag coefficient of a silver dollar varies depending on how it's thrown. If it's tumbling, it has a much higher drag coefficient, but if it's spinning like a Frisbee, it flies more efficiently.

LOW DRAG   HIGH DRAG

(George Washington, silver dollar (tumbling)) → 176 feet
(George Washington, silver dollar (spinning)) → 467 feet

The Rappahannock river is only 372 feet wide at the place where Washington supposedly threw it. With the right spin, it's possible he could have made the throw! (The Potomac, at over 1,800 feet, is too wide.) And, confirming this, many people have successfully recreated the throw. In 1936, retired pitcher Walter Johnson successfully threw a silver dollar 386 feet over the Rappahannock. A day earlier, first baseman Lou Gehrig threw a silver dollar over a 400-foot-wide stretch of the Hudson.

Our model is just an approximation. But the answers it gives don't seem too far from reality, and it's remarkable that we can get even vaguely realistic answers about a complex physical action like "throwing" using so few pieces of elementary physics.

At least, the answers are realistic in some sense, if not in others.

(Carly Rae Jepsen, George Washington) → 35 inches

CHAPTER 11

How to Play Football

There are lots of games called "football," connected through a complicated genealogical tree.

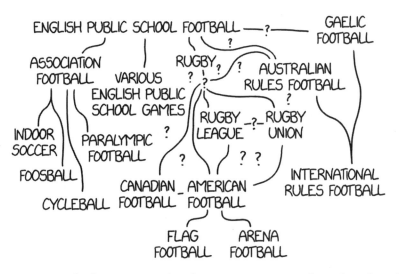

If you're not sure which version you're playing, you can try asking the other players, or watch what people are doing and guess from context.

ALL RIGHT, TEAM: WHICH FOOTBALL ARE WE PLAYING?

I SAW SOMEONE KICKING A BALL.

THAT COULD BE ANY OF THEM.
DID THE BALL LOOK ROUND?

I COULDN'T TELL.

WELL, JUST REMEMBER: IF YOU SEE
BLACK AND WHITE, *NO TACKLING.*

Most versions of football have a number of elements in common. They involve two teams of a dozen or so players, one team on each side of a large field, each trying to get a ball into the goal at the opposing team's side. They also almost always feature kicking at some stage of the game, but different versions allow you to touch the ball with different parts of your body.

There are lots of players on the field, but generally only one of them can have the ball at a time, so there are plenty of opportunities for you to just run around on the field without ever having to deal with the ball. You can just do your best to look busy, and as long as you don't get near the ball, maybe no one will notice you.

Eventually, someone may try to give you the ball – this happens a lot if you're playing American football and you're the quarterback. Or you might get bored with running around and decide to *take* the ball, either by catching it or – depending on the rules – grabbing it from someone as they go by.

EVERYONE IS SO EXCITED TO HOLD THAT BALL. MAYBE I SHOULD SEE WHAT ALL THE FUSS IS ABOUT.

Once you have the ball, *everyone* will pay attention to you, and lots of people will try to take it away. If you aren't enjoying the pressure, you can hand it off to a teammate.

If you're feeling ambitious, you can try to score points yourself. In football, as in so many sports, the general way to accomplish this is simple: get the ball to the goal.

THROW THE BALL INTO THE GOAL

In some types of football, you can score points by launching the ball into the goal from a distance, which you can do by throwing, kicking, or using some other part of your body.

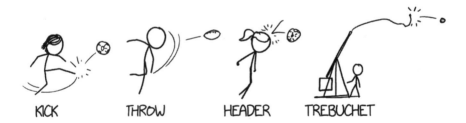

Throwing or kicking the ball directly into the goal might not be an option. In some cases, the rules might not allow a goal to be scored this way—in American football, the quarterback can't just throw the ball through the goal (although it must be tempting sometimes.)

If you decide to attempt to throw or kick the ball into the goal, make note of the distance to the goal and weight of the ball, then turn to chapter 10: How to Throw Things.

In versions of football where you *are* allowed to fling the ball directly into the goal, throwing from a great distance might not be very effective. In association football, for example, it's perfectly legal for the goalie to throw the ball into the opposing team's goal, but it almost never happens. When the goalie tries to throw the ball that far, it generally bounces, rolls, and slows down, giving the opposing goalie plenty of time to catch it.

If you want to try to score a point, but you don't think you can throw the ball from where you are, you'll need to take it to the goal yourself.

TAKE THE BALL TO THE GOAL YOURSELF

Based on the distance alone, walking over to the opposing goal with the ball should only take you a minute or so, and less if you're willing to jog:

But be warned: the other players may not cooperate—particularly the ones on the opposing team.

The other team may try to put players between you and the goal to prevent you from reaching it. Unless you're much larger and stronger than the other players, this will be a problem—and, unfortunately for you, most football teams are made up of people who are both large and strong. You can try running around them, but it's harder than it looks—football players are pretty fast, and they know sometimes people try tricky stuff like that, so they're ready for it.

If another team is trying to stop you from getting to the goal, running faster won't help. The players weigh as much as you, and there are a lot of them, so they'll be able to absorb almost all of your forward momentum. It would take a huge amount of power to push through them.

One way to get through a wall of opposing players would be to take steps to increase your weight, speed, and power.

A person on a very large horse has a combined weight roughly equal to that of an American football team, and the horse's high speed would give a momentum advantage, making it easier to shove through the opposing team.

FIFA's *Laws of the Game*, the official rules for association football, do not contain the word "horse,"[1] so you could try to make an *Air Bud* argument: there's no rule in the books that says you can't use a horse in football. There are rules against *equipment*, but a horse isn't equipment—it's a horse.

The referees may not find your argument convincing. If you ride a horse onto the field, there's a good chance they'll try to stop you. Referees are typically smaller than players, and there aren't as many of them, but they'll still add to the crowd that you have to push through on your way to the goal. They'll likely also decide that your goal doesn't count, but at this point you've probably forfeited that already.

A horse is much bigger than a person, and can certainly knock a handful of people out of its way. But a larger group of people might present too much of a barrier for even a large horse to push through.

The climactic battle at the end of the *Lord of the Rings* movie trilogy showed horses riding through a seemingly endless sea of orcs, knocking them out of the way as they went. Would it be possible for a horse to do this without losing speed?

We can actually answer this question using equations for air resistance—except with orcs in place of air.

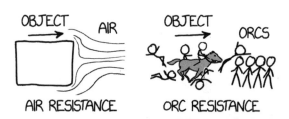

---

1   The NFL rules actually *do* contain the word "horse," but only in reference to a move called a "horse-collar tackle."

The basic formula for calculating air resistance is the drag equation:

$$\text{drag force} = \tfrac{1}{2} \times \text{drag coefficient} \times \text{air density} \times \text{frontal area} \times \text{speed}^2$$

When an object flies through the air, it runs into air molecules and has to push them out of the way. In a sense, the drag equation can be thought of as representing the total mass of air that the projectile has to move through, and how much momentum that air carries:

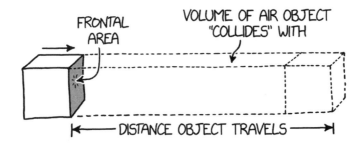

The main parts of the drag equation can be derived from this diagram.[2] When an object goes faster, it collides with more air molecules per second, *and* those air molecules are moving faster relative to the object, which is why speed is squared. If an object's speed doubles, it will hit twice as much air per second *and* the air will be going twice as fast, so the impulse delivered by the air each second – the force – goes up by a factor of 4.

We can use this equation to calculate how much power an object needs to exert to overcome this resistance and maintain its speed. Energy is force times distance, and power is energy per second, so the power the object must exert is equal to the drag force times the distance it travels each second. Since distance per second is speed, therefore power equals drag force times speed. We already multiplied by speed twice to get drag force, and now we need to multiply by speed *again*:

$$\text{power} = \tfrac{1}{2} \times \text{drag coefficient} \times \text{air density} \times \text{frontal area} \times \text{speed}^3$$

---

2   If you've taken some physics classes and you stare at this diagram long enough, you might start to wonder what the ½ is doing in the drag equation. Since the drag coefficient is a unitless, arbitrary scale factor, the ½ could be eliminated just by doubling all the drag coefficients. Sports physicist John Eric Goff has pointed out that if you derive the equation by thinking about the momentum carried by the incoming air molecules, it seems like a factor of 1 – or possibly 2 – would be more natural than ½. However, if you think of drag in terms of the kinetic energy of the incoming air, then carrying over the ½ from the kinetic energy equation makes more sense. Physicists tend to explain it this way – by saying the drag equation represents the "dynamic pressure" of the incoming air – but not every authority agrees. Frank White's *Fluid Mechanics* textbook simply calls the factor of ½ a "traditional tribute to Euler and Bernoulli."

That "3" in the exponent tells us that as an object goes faster, the power it has to exert to overcome drag rises very quickly.

Strangely enough, we can use this same approach to estimate how much energy a horse would exert while plowing through a crowd of orcs, by treating the orcs as a uniform gas with very large molecules.

Adapting the equation to the horse-orc geometry gives us this equation for power:[3]

$$\text{power} = \text{orc crowd density} \times \text{orc weight} \times \text{horse chest width} \times \text{speed}^3$$

Note: We've gotten rid of the factor of ½ and the drag coefficient. For a "gas" made up of individual non-interacting molecules that bounce off the front of a curved object as it moves, the drag coefficient works out to around 2.

The orcs in the film are probably standing with a density of roughly 1 orc per square meter. If we assume each orc weighs 200 lb, and the horse is 2½ feet wide at the chest and galloping at 25 mph, that works out to:

$$\frac{1 \text{ orc}}{\text{m}^2} \times \frac{200 \text{ lb}}{\text{orc}} \times 2.5 \text{ feet} \times 25 \text{ mph}^3 = 97 \text{ kilowatts}$$

Can a horse sustain a power output of nearly 100 kilowatts? To know that, we need to know the sustained power output of a horse. Conveniently for us, "horsepower" already exists as a unit, so the calculation is just a simple unit conversion:

$$97 \text{ kilowatts} \approx 130 \text{ horsepower}$$

One hundred thirty horsepower is too much for one horse. A horse can do more than 1 horsepower of work for a short time – a horsepower is defined by work done over a long period – but a horse's maximum short-term output is more like 10 or 20 horsepower, far

---

3   This horse-drag equation doesn't have a common name in physics, and honestly it would be pretty weird if it did.

short of the 130 needed for the feat in the movie. To reduce the power required to push through the crowd of orcs, the horse would need to slow to a trot.

The orc resistance equation also applies to football players, referees, and any other horde of enemies that you might want to charge through on horseback. If you're going to push through a crowd of players on horseback, you'll need to slow way down, which will give your opponents a chance to brace against you, climb onto your horse to overload it, or grab your legs and drag you from the saddle onto the field where you can be tackled in the traditional manner.

Like any trick play, the horse gambit loses effectiveness if the opposing side has a chance to prepare for it. Once the opposing players get wind of your plan, they'll be able to prepare with anti-horse defensive measures, like long spears braced in the ground, trenches dug in the field, or strategically placed treats to distract your mount.

But with only a handful of players on the field, if you aim for an opening in their defensive line, you might be able to make it through with just a few collisions. No human runner can catch up to a horse at full gallop, so once you get past the defenders, you'll have a clear shot all the way to the goal.

CHAPTER 12

How to Predict the Weather

What will the weather be like tomorrow?

When people talk about the weather in their particular location, they often repeat an old saying: *"If you don't like the weather in [insert location here], just wait five minutes."* Like every clever saying, it's often attributed to Mark Twain. In this case, he probably *did* actually say it, but if it turns out he didn't, you can just attribute it to Dorothy Parker or Oscar Wilde.

People repeat this quote just about everywhere in the temperate zones, because weather changes all the time and we're constantly surprised by it for some reason.[1] These changes can be hard to predict, but since weather is something that everyone has to deal with—we're all trapped together at the bottom of this atmosphere—we try anyway.

There are lots of ways to predict weather, some better than others. The best modern weather prediction involves sophisticated computer models, but let's start with a basic, time-honored technique: *guessing at random.*

---

1   We humans are good at being surprised by predictable changes. Every time I see a friend with their baby, I feel the urge to comment, "Oh, you've grown since I last saw you!" Apparently some part of me expects babies to stay the same size or get smaller over time.

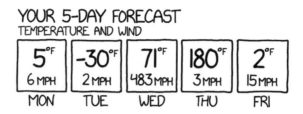

This doesn't work very well at all.

A slightly better method is to make up your prediction by looking at the average weather for that place at that time of year. This is called *climatology forecasting*.

In places where the weather doesn't change very much, like the tropics, this is a pretty good method. For example, the average high temperature in Honolulu, Hawaii in mid-July is 88°F, so we can use that to make a forecast for next July:

Here's the actual recorded temperature on those days for a recent year—2017—in Hawaii:

HONOLULU
ACTUAL HIGH TEMPERATURE

88°F	88°F	87°F	88°F	88°F	89°F	87°F
JUL 13	JUL 14	JUL 15	JUL 16	JUL 17	JUL 18	JUL 19

Nice! Our "prediction" holds up pretty well. We nailed the temperature exactly on 4 of the 7 days, and were never off by more than a degree. Fame and fortune as a weather forecaster lie in our future.

Now let's take this great method and apply it to Saint Louis, Missouri in September. The average high in mid-September is 79°F, so we'll use that to make our forecast:

ST. LOUIS
7-DAY FORECAST

79°F	79°F	79°F	79°F	79°F	79°F	79°F
SEP 13	SEP 14	SEP 15	SEP 16	SEP 17	SEP 18	SEP 19

Here's the actual 2017 temperature on those days:

ST. LOUIS
ACTUAL HIGH TEMPERATURE

76°F	88°F	90°F	91°F	82°F	85°F	89°F
SEP 13	SEP 14	SEP 15	SEP 16	SEP 17	SEP 18	SEP 19

Yikes. We were *way* off!

Forecasting based on averages works better in the tropics because there's less variation in the weather. In the temperate zones, where St. Louis is located,[2] weather is dominated by the movement of big, slow high- and low-pressure systems, which can lead to heat waves, cold snaps, and lots of complaining.

---

2   As of 2019.

Overall, guessing based on averages seems like a bad strategy. But before we move on to good strategies, there's another bad strategy we might want to consider: looking at the weather right now and assuming it will never change.

This sounds silly, because weather changes constantly, but it doesn't change *that* fast. If it's raining now, it's pretty likely to be raining 30 seconds from now. If it's unusually hot right now, there's a decent chance that it will still be unusually hot an hour from now. You can use this principle to make a forecast—just check the weather right now. That's your forecast. This is called *persistence forecasting*.

Over very short time ranges, persistence forecasting works better than forecasting based on averages, and over very long time ranges, average forecasting works better. In some areas of the world, where weather patterns tend to linger for days at a time, persistence forecasting is more useful. In other areas, the weather one day has almost no connection to what the weather will be like the next day. In those areas, forecasting based on averages works better.

COMPUTERS

In the years following World War II, at the dawn of the computer era, the mathematician John von Neumann launched a project to use computers for weather forecasting. By 1956, he had concluded that forecasting could be divided up into three domains: the short term, the medium term, and the long term. He correctly figured out that the approaches needed in all three would be very different, and that the middle domain—the medium term—would be the hardest.

The short term covers the next few hours or days. Over this range, predicting the weather is a matter of getting enough data and then doing lots of math on it. The atmosphere operates according to relatively well-understood laws of fluid dynamics. If you can measure the atmosphere's current state, you can run a simulation which shows how it will evolve. These simulations will give us pretty good forecasts over the next few days.

We can improve these forecasts by gathering more information about the state of the atmosphere, combining feeds from weather balloons, weather stations, aircraft, and ocean buoys. We can also improve the simulations, using more computing power to run them at higher and higher resolution.

But when we try to extend the forecast into the range of several weeks, we run into a problem.

Edward Lorenz, working on computer weather prediction in 1961, noticed that when he ran two versions of a simulation with a very tiny difference between them – like adjusting the temperature at one location from 50°F to 50.001°F – the outcome would be totally different. The disparity wouldn't be noticeable at first, but gradually, the small difference would grow and spread outward through the system. Eventually, the systems would look nothing like each other on the large scale. He coined the term *butterfly effect* for this – based on the idea that a butterfly flapping its wings on one side of the world could eventually change the course of storms on the other side of the side of the planet. This idea developed into chaos theory.[3]

Because weather is a chaotic system, the medium-term forecast – what will the weather be like a month or a year from now – may to some extent be fundamentally unknowable. We've discovered some slow-moving cycles that drive seasonal changes, like El Niño and the Pacific Decadal Oscillation, which give us hints about overall attributes of the next season. But it may never be possible to predict on May 1 whether it will rain on October 1.

The long-term domain covers a range of decades to centuries, and is what we now think of as climate change prediction. Over long-time horizons, the chaotic day-to-day variation averages out, and the climate is dominated by long-term energy inputs and outputs. It's probably never possible to make perfect climate predictions – since the underlying chaos can always throw a wrench into the system – but we can say with some confidence how things will change, on average. If the amount of sunlight entering the atmosphere goes up, so will the average temperature. If the amount of CO_2 in the atmosphere goes down, more infrared radiation escapes the surface, and the temperature goes down. There are all kinds of complicated feedback loops involved, some of which we don't yet fully understand, but the basic behavior of the system is in principle predictable.

Here's where that leaves our three domains:

- **Short:** fully predictable, with good-enough computer simulations
- **Long:** hard to predict with certainty, but possible on average
- **Medium:** may be literally impossible

People used to complain all the time about weather forecasts being wrong. They still do, of course, but the complaints may be getting a little less common. As we improve our computer simulations and our data gathering, our predictions over the short term – the ones that go into the 5-day weather forecast – are getting steadily more accurate. By 2015, the 5-day forecasts were as accurate as 3-day forecasts in 1995. In the mid-20th century, forecasts of the weather more than two or three days away were no better than the fore-

---

3   And, according to *Jurassic Park,* somehow led to a bunch of dinosaurs eating people.

casts you'd get from the simple persistence and averaging methods—which don't take any computers at all. Now, our best computer models are making weather forecasts that outperform those simple methods up to 9 or 10 days ahead of time.

In general, over the last half century, weather forecasts have been improving at a rate of 1 day per decade, which works out to about 1 second per hour.[4] Physics calculations suggest that the fundamental limit to our simulation-based forecasts is probably in the range of a couple of weeks. After two or three weeks, the inherent chaotic nature of the system renders prediction impossible.

But you don't necessarily need access to a supercomputer to predict the weather.

RED SKY AT NIGHT

According to popular lore, you can predict the weather based on sky color. The saying typically goes, *"Red sky at night, sailor's delight. Red sky at morning, sailors take warning."*

This saying has been around in various forms for a long time—there's a version of it in the Bible.[5] The reason it's lasted so long is that it actually works, at least in certain parts of the world. The red sky method doesn't have as much to do with the red clouds themselves as you might think—instead, it's a way of using the Sun to do an X-ray of the atmosphere over the horizon, and then using clouds above you as a screen on which to project the results!

WAIT, WHAT?

In the temperate zones, weather systems generally move from west to east. They don't move too fast—in general, weather moves across the Earth at driving speed or slower—so a storm system a thousand miles to your west won't reach you for a day or so. Because of the curve of the Earth and the haze of the atmosphere, you can't see the clouds to your west; if you could, weather forecasting would be a lot easier.

---

4   If you want to annoy a physicist, mention to them that the SI unit for "seconds per hour" is "radians."

5   *"When it is evening, ye say, 'It will be fair weather: for the sky is red.' And in the morning, 'It will be foul weather today: for the sky is red and lowering.'"*- Matthew 16:2–3.

The "red sky" trick gets around this by using the Sun. Red wavelengths pass through air more easily than blue ones. When the Sun is setting in the west, its light passes through hundreds of miles of atmosphere—becoming extremely red in the process—before hitting the clouds above you. Shorter blue wavelengths bounce off the air and go off in other directions. This is why the sky is blue—it reflects blue light. White clouds reflect all colors, so when red light shines on them, they look red, too.

If there are storm clouds to your west, the red sunlight is stopped before it can get to you, and the sunset doesn't look particularly red:

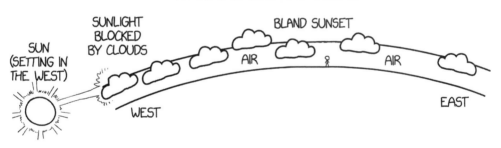

On the other hand, if there's clear air for hundreds of miles to your east, the sunlight passes all the way through to reach the sky above you, turning it red. If there are any clouds overhead, the red light illuminates them, creating a spectacular sunrise.

RED SKY AT MORNING

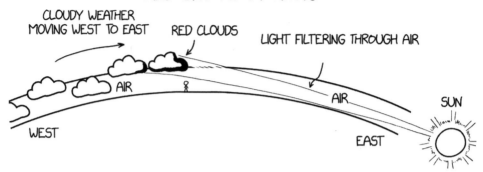

When weather moves west to east, a red sky at night means that there are clouds overhead, but clear skies to the west – which tells you that the weather will likely be clearing up.

A red sky in the morning, on the other hand, means that there's clear air to the *east* . . . but clouds overhead. That means the clear zone is moving away, and clouds are moving in.

This saying doesn't work in the tropics, where prevailing winds tend to move east to west and are generally more unpredictable. On the other hand, weather in the tropics is much more stable—excepting the occasional unpredictable cyclone—so there's less need for this kind of rule of thumb.

GOLDEN HOUR

The filtering effect of the atmosphere is part of why the time near sunrise and sunset is referred to as the "golden hour" in photography. The same warmer, redder light that creates brilliant sunsets makes for good portraits, as well as good sunset photos.

This means that in temperate zones, you can get a hint of the coming weather just by looking at the photos being posted online. If you check Facebook in the evening and see sunset photos with a higher-than-normal number of red and yellow pixels, and warmly lit selfies getting an unusual number of likes, it suggests bad weather is leaving the area. Sunrise photos and glowing morning selfies, on the other hand, may be an ominous sign.

The photo-color forecast may not be as reliable as a supercomputer-simulation of the atmosphere, but it's pretty impressive for an ancient method contained in a catchy rhyme.

If you're not a sailor, you can tweak the wording as needed.

How to
Go Places

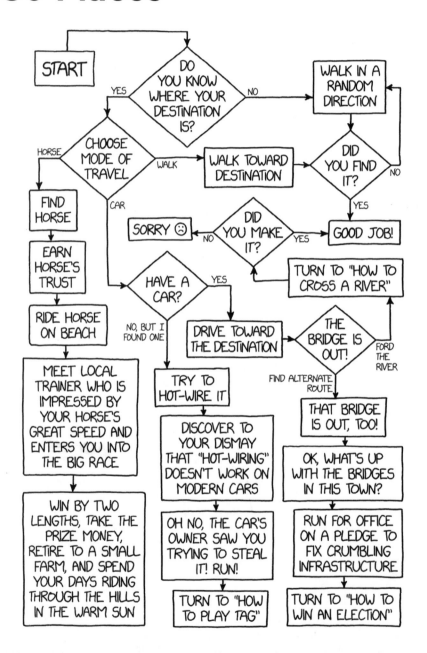

CHAPTER 13

How to Play Tag

The rules of tag are simple: One player is **It**, and tries to chase and touch another player. If **It** catches someone, that person becomes **It**.

There are countless variations on the basic rules of tag—there's even a parkour-like league called World Chase Tag that hosts competitions in which athletes chase each other while leaping and diving over obstacles—but the standard playground version of tag has very few specific rules. It doesn't require scores, goals, equipment, or a defined playing area. Playground tag typically doesn't even have a defined end. You can't win at tag; you can only stop playing.

In theory, in an idealized game of tag—in which some players are faster than others, and everyone runs at their top speed—the action should eventually reach a natural equilibrium. If the person who is **It** is not the slowest player, then they will catch a slower player, tag them, and no longer be **It**. But eventually the slowest player will become **It**, will be unable to catch and tag another player, and will remain **It** forever.

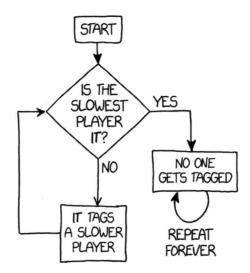

If the game never ends, the non-**It** players have to keep running. If they stop to rest, they risk a tortoise-and-the-hare scenario. If you're a player who's faster than **It**, but you want to sleep eight hours every day, you have to make sure you get a large enough lead over your opponent that you can rest without them catching up.

Our model is still highly idealized. In reality, runners don't just have a "top speed." Some people are fast over short distances, while others can maintain a pace for a long time. Adding this to our simplistic tag model makes it a little more interesting.

Let's imagine a game of tag between Usain Bolt—the world's fastest short-distance sprinter—and Hicham El Guerrouj, who holds the world record for the mile run. We'll assume both runners are in their athletic prime, and use measurements from their world-record races to model their pace.

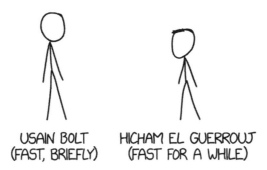

USAIN BOLT
(FAST, BRIEFLY)

HICHAM EL GUERROUJ
(FAST FOR A WHILE)

Long-distance runners and sprinters rely on different physiological mechanisms for power. A sprinter relies on anaerobic processes, which provide a lot of energy over short distances, but after a minute or two exhaust the body's energy reserve. Long-distance runners rely on more aerobic, oxygen-consuming processes, which provide a steadier supply of energy over long distances.

Usain Bolt is the current world record holder in most sprinting events. He's the fastest person in the world… unless you need to run farther than a few hundred meters. His time in the 400-meter sprint is good, but more than two seconds behind the world record.[1] At distances beyond that, he doesn't even match a good high school sprinter. Bolt's agent told *The New Yorker* that Bolt has *never* run a mile.

Let's assume the game of tag starts with El Guerrouj as **It**, although it doesn't actually matter who's **It** at the start—if Bolt is **It**, he'll simply sprint forward and tag El Guerrouj within the first few seconds.

BOLT IS IT   EL GUERROUJ IS IT

Bolt, to escape being tagged, would start to run. At first, he would have the advantage; his sprinting ability would let him quickly put distance between himself and the slower

---

1   400 meters is just barely long enough to use up a sprinter's anaerobic reserves and require some aerobic energy.

El Guerrouj. Thirty seconds into the game, as Bolt passed 300 meters, he would be a full 70 meters ahead of his pursuer.

After 30 seconds, however, the gap between them would start to close. A little over 90 seconds into the game, El Guerrouj would catch up to Bolt just short of the 700-meter mark and tag him.

Bolt, exhausted, could try to chase him, but would be unable to catch up.

Assuming you're not a marathon champion yourself, good long-distance runners will have a huge advantage over you in games of tag. Whether you're Usain Bolt, Uwe Boll,[2] Ugo Boncompagni,[3] or *Usnea barbata*,[4] you won't be able to catch a marathon runner once they get up to speed.

If you do find yourself in Bolt's shoes, facing someone with superior distance-running skills, are you doomed to be **It** forever?

Well, maybe.

---

2   A horror filmmaker
3   The birth name of Pope Gregory XIII
4   A species of lichen

HOW TO CATCH A LONG-DISTANCE RUNNER

If you can't catch a runner by running, you can try the more efficient option: walking.

Walking is slower than running, but it's substantially more energetically efficient—it requires less oxygen and fewer calories per mile. This is why a healthy person might struggle to run a mile, yet be able to walk for several hours without any serious difficulty. Running puts a higher demand on your aerobic system; if your body can't keep up with that demand, you can't keep running. Distance runners learn to run in ways that waste as little energy as possible, but they also condition their cardiovascular system to supply energy at a rate that can meet the demand of sustained running.

Hikers typically walk the 2,190-mile Appalachian Trail in 5 to 7 months. The lower number works out to a pace of a little under 15 miles per day, so let's assume you can keep up a 15-mile-per-day pace indefinitely.

Yiannis Kouros, a long-distance running champion, once ran 180 miles in a single 24-hour period. If you were chasing Kouros at a hiker's pace, he could spend the first day running 100 miles to get away from you; then he could rest for another week or so as you caught up. Once you got close, he could run another hundred miles.

If Kouros wanted to live a normal life – but was determined not to be tagged – he could buy two or three houses 100 miles or so apart. When you got close to one of them, he could run to the next one. That way, he could have a week or so of rest at each house before you caught up and forced him to flee to the next house.

Hopefully, any family he lives with are also marathon runners – otherwise, he'll have to do a lot more work to stay ahead of you.

HOW TO *ESCAPE* A MARATHON CHAMPION

If you finally manage to sneak up on Kouros and tag him while he's not paying attention, you'll have a new problem: He's just going to immediately tag you back. You certainly won't be able to outrun him.

If you can't win within the rules, maybe you can bend them slightly. Let's suppose you hop on a magic scooter, one which lets you "run" as fast as you want, and tag Kourous immediately.

Let's say Kouros, once he becomes your pursuer, refuses to stoop to your level, and insists on chasing you the old-fashioned way. No matter how far you go, he'll keep chasing you – but if you can get really far away, you can give yourself lots of time to rest and relax.

You can use Google's walking directions to try to find the two points on Earth with the longest walking distance between them. The points change over time as Google updates its maps, but science artist Martin Krzywinski has collected a list of them. One promising candidate is a trip from Quoin Point, in South Africa, to Magadan, a city on the eastern coast of Russia.

This route is about 14,000 miles long. It crosses through 16 countries, involves ferries over a number of rivers and canals,[5] and makes more than two dozen total border crossings. All in all, the walking itinerary includes roughly 2,000 individual directions.

"TURN RIGHT.
WALK 15.8 MILES.
CONTINUE ONTO RT. B8.
GO 1.2 MILES.
ENTER TANZANIA.
CONTINUE..."

The route is hilly—over a hundred kilometers of total elevation change—and crosses through practically every climate zone, from tropical rain forest to hot desert to Siberian tundra. It's hard to say how fast your pursuer will be able to travel it, but the record time

---

5   Depending on road closures and border procedures, you may also need to take a ferry on Lake Nubia/Lake Nasser in order to cross the Egypt/Sudan border.

for hiking the Appalachian Trail is currently a little over 41 days, an average of 53 miles per day. At that pace, the trek between Quoin Point and Magadan will take about nine months.

You can continue going back and forth indefinitely, uprooting your life every year or so to move across the world, until your pursuer gives up.

Or you can talk things over. If a game of tag never ends, and someone has to be **It**, why not share? Rather than running back and forth across the world, you could just pick a nice place to live – perhaps some town you've found on your travels. You and your fellow play-ers could move into houses right next to each other, and trade off **It** status every day...

... by sharing a daily high five with your new neighbors.

Maybe there *is* a way to win tag after all.

CHAPTER 14

How to Ski

Skiing involves strapping long, flat objects to your feet and sliding across a surface or down a slope. The surface is usually water, either in frozen or liquid form, but it doesn't have to be.

You can slide down any slope if it's steep enough. When an object sits on a slope, gravity partly pulls it downward and partly pulls it along the slope. An object starts to slide when the force pulling it along the surface overcomes the force of friction.

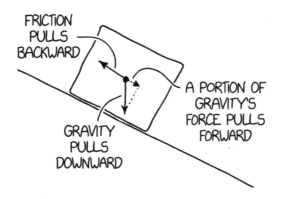

Depending on what material your skis and the surface are made of, you might not start sliding easily. If the skis are made of rubber, and the surface is cement, you'll need quite a steep slope to ski, which is presumably why rubber-on-cement skiing is so unpopular.[1]

For any combination of surface material and ski material, you can use a simple physics relation to calculate how steep the slope will have to be to slide. It seems like it might be a hard problem, but thanks to a convenient coincidence, most of the complicated parts cancel out, and you wind up with this extremely straightforward equation:

$$\text{coefficient of friction} = \tan(\text{slope angle})$$

If you want to know the slope angle, you can reverse the equation:

$$\text{slope angle} = \tan^{-1}(\text{coefficient of friction})$$

This equation is delightfully uncomplicated, right up there with $E=mc^{2}$[2] and $F=ma$. Unlike those more famous equations, it's only useful for this very specific problem, but it's still neat how simple it is.

Here's a table of coefficients of friction of different ski/surface materials:

Ski material

Surface	Rubber	Wood	Steel
Concrete	0.90	0.62	0.57
Wood	0.80	0.42	0.3
Steel	0.70	0.3	0.74
Rubber	1.15	0.80	0.70
Ice	0.15	0.05	0.03

Here's a table of coefficients of friction and the corresponding minimum slope angle you need to start sliding:

- **0.01/0.6°** (bicycle on wheels)[3]
- **0.05/3°** (teflon on steel, ski sliding on snow)
- **0.1/6°** (diamond on diamond)
- **0.2/11°** (plastic shopping bags on steel)
- **0.3/17°** (steel on wood)

---

1   Ironically, it's never really gained traction.

2   The second 2 is a footnote, not a superscript.

3   Bicycles have wheels, but they're still subject to friction – the wheels just move the location of some of the friction from the ground to the bearings of the axle.

- **0.4/22°** (wood on wood)
- **0.7/35°** (rubber on steel)
- **0.9/42°** (Rubber on concrete)

Wooden skis would work on a 16° steel ramp. If the skis were made of rubber, a steel ramp would need to be 35° before you could slide. The coefficient of friction between rubber and concrete is even higher – 0.9 – and you'd need a pretty steep slope of about 42° in order to slide down. This also tells you that a person in rubber-soled sneakers can't walk up a ramp with a slope steeper than 42°.

In a sense, skiers are really just mountain climbers who are unusually bad at climbing but make up for it with very good balance.

Ice is slippery compared with most surfaces, and snow – which is really just fancy ice – is similarly slick. This makes them a good choice for skiing and similar activities, which is why every sport in the Winter Olympics involves sliding in some way.

The reasons that ice is slippery are actually a little mysterious. For a long time, people believed that the pressure from a skate blade melted the surface of the ice to create a thin, slippery layer of water. Scientists and engineers in the late 1800s demonstrated that the pressure of an ice skate blade could lower the melting point of ice from 0°C to −3.5°C. For decades, pressure melting was accepted as the standard explanation for how ice skates work. For some reason, no one pointed out that it was possible to skate at temperatures colder than −3.5°C. The pressure melting theory suggests it should be impossible, but ice skaters do it all the time.

The actual explanation for why ice is slippery is, surprisingly, still the subject of ongoing physics research. The general explanation seems to be that there's a layer of liquid water on the surface of ice because the water molecules aren't firmly locked into the ice crystal lattice. In this way, an ice cube is sort of like a piece of cloth with fraying edges. In the middle of the cloth, threads are locked into a well-organized form, but on the edges, they're less constrained and more likely to come loose and move around. In the same way, water molecules near the edge of a piece of ice come loose and move around, creating a thin layer of water. However, the properties of this water layer and how a skate interacts with it aren't fully understood.

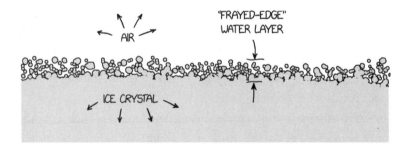

Given how much time modern physics spends on deep and abstract mysteries like searching for gravitational waves or the Higgs boson, it can be surprising how many basic everyday phenomena aren't well understood. In addition to ice skates, physicists don't really understand what causes electric charges to build up in thunderstorms, why sand in an hourglass flows at the speed it does, or why your hair gets a static charge when you rub it with a balloon. Fortunately, skiers and skaters can slide on snow and ice without waiting for physicists to finish figuring things out.

Snow is already pretty slippery, but to gain a little extra slickness, skiers add a layer of wax to their skis. The wax serves as a semiliquid layer, keeping sharp ice crystals from digging into the hard material of the skis and slowing them down.

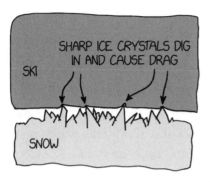 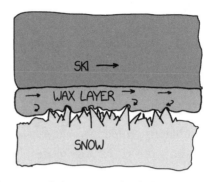

Waxed skis on snow have a coefficient of friction of about 0.1, which drops to 0.05 once the skis start moving.[4] This means that you need a slope of 5° to start sliding under your own weight, but once you get moving, you only need a slope of about 3° in order to keep going.

Once you're sliding down a slope, you'll continue accelerating until you either run out of snow or you reach a speed at which air resistance is pulling backward harder than gravity is pulling you forward. Since air resistance doesn't really start to kick in until higher speeds, even a gentle slope can let a skier or sledder go pretty fast if it's long enough. The theoretical top speed of a skier or sledder on a 5° slope of unlimited length is around 30 miles per hour—45 if they're particularly aerodynamic. On a 25° slope, speeds of over 100 mph should be possible for an aerodynamic skier or sledder.

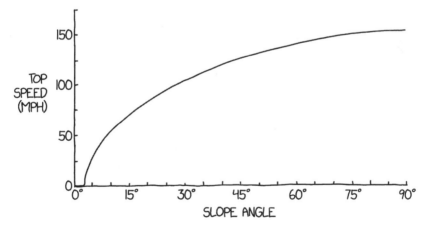

The world record for top speed achieved on skis is around 155 mph, but people don't keep close track of that record, because it turns out not to be a particularly interesting boundary to push. The way to reach higher speed is simply to find a longer, steeper slope.

---

4   Objects have a lower coefficient of friction when they start moving. This is the reason why, when you slip on a patch of ice, your feet go out from under you so abruptly. As soon as your shoes start to move, they lose their grip completely.

If you keep doing that, skiing gradually morphs into skydiving—only an even more dangerous version of skydiving, since instead of falling through open air, participants are skimming across ground. Obstacles are very hard to avoid when skiing at 155 mph, and even if you find what seems like a smooth slope, a small bump or gentle turn could be instantly fatal.

When a competitor's score in a sport is strongly correlated with their odds of dying, it creates obvious problems for the sport. Speed skiing was briefly featured at the 1992 Olympics, but, after a number of deadly accidents, has been mostly abandoned at the competitive level.

WHEN YOU REACH THE BOTTOM

If you're skiing down a slope, eventually you'll reach a point where you can't go forward any more. This can happen for a few reasons:

- There are trees, rocks, or hills in the way
- You reached the bottom of the mountain
- There's no more snow

If you're having fun and don't want to stop skiing, you have a few options.

If there are trees in the way, you can try removing them; for more on how to do this, turn to chapter 25: How to Decorate a Tree. If there are rocks in the way, you can check chapter 10: How to Throw Things for advice on whether you can move them. If you've reached the bottom of the mountain, you can try continuing to accelerate yourself forward; you may find some helpful advice in chapter 26: How to Get Somewhere Fast or chapter 13: How to Play Tag. If you want to continue downhill even though there's no more hill, turn to chapter 3: How to Dig a Hole.

If you've run out of snow, read on.

WHAT TO DO IF YOU RUN OUT OF SNOW

From our discussion of friction, we know skis don't work very well on most non-snow surfaces. There are some artificial ski slopes that use special low-friction polymers, with a bristly hairbrush-like texture that provides some softness and lets the skis dig in when turning. There are also special skis designed for use on grass and other surfaces, but they use wheels or treads rather than sliding.

If you want to keep skiing on snow, but there's no more snow to slide on, you'll have to make some yourself.

About 90 percent of American ski resorts use artificial snow to ensure that ski slopes are covered as soon as it's cold enough for snow to stick, and to keep them covered for the whole ski season even if the weather doesn't cooperate. The artificial snow also helps to replenish snow lost throughout the season due to melting and erosion from skiers.

Snow machines make artificial snow by using compressed air and water to create a stream of tiny ice crystals, and then misting the ice crystals with more water droplets as they float in the air. As the mist drifts down to the ground, the water droplets freeze onto the ice crystals to form snowflakes.

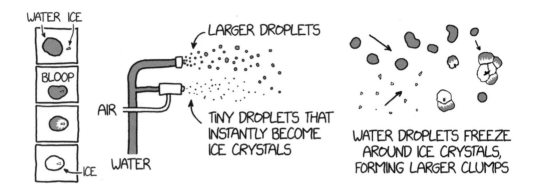

The snowflakes formed in this way are more compact and misshapen than the delicate shape of natural snowflakes. Natural snowflakes have much more time to grow slowly in a cloud, one water molecule at a time, which allows intricate and symmetrical shapes to form. Artificial snow forms quickly, in the short time it takes water to descend from the nozzle to the ground, from a handful of drops clumsily jumbled together.

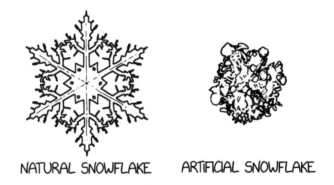

NATURAL SNOWFLAKE   ARTIFICIAL SNOWFLAKE

Suppose you need a 5-foot-wide path to ski on, and you're going to descend at a speed of 20 mph. Natural snow might be 10 percent water and 90 percent air by volume, although this ratio varies quite a bit depending on how light and fluffy the snow is. For simplicity's sake, let's also assume you want about 8 inches of somewhat heavy snow to ski on, with snow that's as 1/8 as dense as water, equivalent in mass to a layer of water an inch thick. The total amount of water you'll need is therefore:

$$5 \text{ feet} \times 8 \text{ inches} \times \tfrac{1}{8} \times 20 \text{ mph} = 90 \ \tfrac{\text{gallons}}{\text{second}} = 1{,}250 \ \tfrac{\text{m}^3}{\text{hour}}$$

Skiing the length of a football field will require a thousand gallons of water, along with the equipment to turn it to snow.

You'll have a hard time finding equipment that can produce snow fast enough for you. The biggest snowmaking machines might produce snow at rates of 100 cubic meters per hour. That's just 10 percent of what you need, so you may need a lot of them.

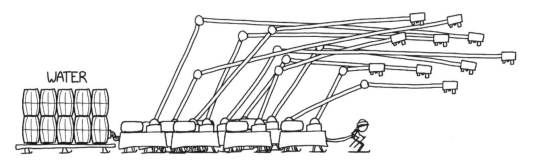

Snow from typical snowmaking equipment needs a lot of time to drift down to the ground, which means you'll have to produce the snow far ahead of your current position to give it time to settle, and the movement of air currents may make it hard to concentrate enough of it along a narrow path.

The long, slow descent is necessary because it takes a long time for the water droplets to lose heat to the air through evaporation to attach to the ice crystals. There *are* ways to cool the water droplets down more quickly – but they have some drawbacks.

If you inject low-temperature substances like liquid nitrogen into the air/water stream, they can reduce the temperature and cause almost-instant freezing. These techniques can produce snow quickly, and are used by some snowmaking companies for special events in areas where the air temperature is too high for normal artificial snow to be produced. Cryogenic freezing techniques are generally *not* used by ski resorts—it's far too expensive and energy intensive to freeze water this way compared to letting it freeze on its own in the air.

For your small, very narrow ski slope, liquid nitrogen just might be affordable. If you buy the liquid nitrogen in the form of small tanks, your ride could cost $50 per second, but industrial suppliers can get you a much better price if you buy in bulk.

You don't have to use liquid nitrogen—you could also try other cryogenic gases. Liquid oxygen is similar to liquid nitrogen and just as easy to produce, and could in theory be used for snowmaking. However, this is not recommended. Liquid nitrogen is a popular cryogenic fluid in part because it's so inert and nonreactive. Liquid oxygen is neither of those things.

MAKE THE PROCESS MORE EFFICIENT

You could reduce the snow consumption if you could somehow scoop up the snow behind you and reuse it, rather than producing more snow as you go.

If you lay down some kind of tarp under the snow, you can pick the whole sheet of snow back up and reuse it with minimal losses.

The tighter you make the snow-transfer loop, the less snow you'll need.

You can even make the loops smaller than your body if you pass the stream of snow around your legs, rather than over your head . . .

. . . at which point you'll realize you've effectively reinvented roller skates.

CHAPTER 15

How to Mail a Package

(from space)

Based on the 2001–2018 average, 1 out of every 1.5 billion humans is in space at any given time, most of them on board the International Space Station.

ISS crew members ferry packages down from the station by putting them in the spacecraft carrying crew back to Earth. But if there's no planned departure for Earth any time soon—or if NASA gets sick of delivering your internet shopping returns—you might have to take matters into your own hands.

THE RETURN LABEL SAYS IT CAN
BE MAILED FROM ANYWHERE!

YOU SHOULD'VE TRIED THE SHOES ON
BEFORE YOU BROUGHT THEM UP HERE.

Getting an object down to Earth from the International Space Station is easy: you can just toss it out the door and wait. Eventually, it will fall to Earth.

There's a very small amount of atmosphere at the ISS's altitude. It's not much, but it's enough to produce a tiny but measurable amount of drag. This drag sooner or later causes objects to slow down, fall into a lower and lower orbit, and eventually hit the atmosphere and (usually) burn up. The ISS also feels this drag; it uses thrusters to compensate, periodically boosting itself up into a higher orbit to make up for lost altitude. If it didn't, its orbit would gradually decay until it fell back to Earth.

Astronauts accidentally deliver packages to Earth this way all the time. While working on the ISS, spacewalkers have accidentally dropped a variety of random objects, including a pair of pliers, a camera, a tool bag, and a spatula an astronaut was using to apply a repair adhesive for testing. Each of these inadvertently created satellites circled the Earth for a few months or years before its orbit decayed.

WHOOPS!
UH, MISSION CONTROL, THIS IS EAGLE ONE.
I'M, UH, PLEASED TO ANNOUNCE THE LAUNCH OF OUR NEWEST SATELLITE.

A package you toss out the door will suffer the same fate as all the lost parts, bags, and random pieces of equipment that have drifted away from the station over the years: it'll deorbit and enter the atmosphere.

ORBITAL DELIVERY

SHIPPING OPTION	SHIPPING TIME	PRICE
◯ EXPEDITED (BALLISTIC DELIVERY)	45 MINUTES	$70,000,000
◯ PRIORITY (SOYUZ DELIVERY + AIRMAIL)	3-5 DAYS	$200,000
◉ ECONOMY (ATMOSPHERIC DRAG)	3-6 MONTHS	FREE

This shipping method has two big problems: First, your package will burn up in the atmosphere before it ever reaches the ground. And second, if it does survive, you'll have no way to know where it will land. To deliver your package, you'll have to solve both these problems.

First, let's look at how to get your package to the ground intact.

REENTRY HEATING

When stuff enters the atmosphere, it often burns up. This isn't because of some weird property of space. It's because everything in orbit is going so *fast*. When objects hit the air at those speeds, the air doesn't have time to flow out of the way. It compresses, heats up, turns to plasma, and often melts or vaporizes the object in the process.

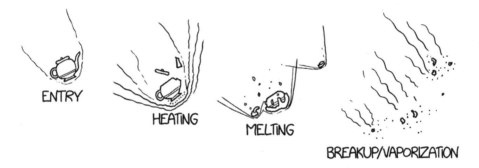

ENTRY HEATING MELTING BREAKUP/VAPORIZATION

To keep our spacecraft from being destroyed, we attach heat shields to the front, to absorb the heat from reentry and protect the rest of the craft.[1] We also give them special shapes, which helps create a cushion of air between the shock wave and the surface of the spacecraft, keeping the hottest plasma from touching the hull.

---

1   Why don't spacecraft slow down using rockets, then enter the atmosphere at a low speed, to avoid the need for a bulky heat shield? The answer is simple: it would take way too much fuel. The spacecraft that use rockets for landing, like the *Curiosity* rover or SpaceX's reusable launchers, do most of their decelerating using atmospheric drag, and only use rockets for the last bit of the landing.
Getting a spacecraft going against gravity fast enough to get into orbit takes dozens of times the spacecraft's own weight in fuel, which is why rockets are so big. Slowing down would take roughly the same amount. Which means that instead of launching a 1-ton spacecraft to orbit using 20 tons of fuel, you'd need to launch a 1-ton spacecraft *and* 20 tons of fuel to slow it back down. But now instead of a 1-ton spacecraft, you're effectively launching a 21-ton spacecraft, which means you'll need 420 tons of fuel. Compared to 420 tons of fuel, a 100-pound heat shield is a *way* more efficient solution.

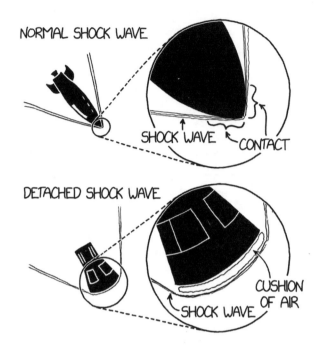

NORMAL SHOCK WAVE

SHOCK WAVE ← CONTACT

DETACHED SHOCK WAVE

SHOCK WAVE

CUSHION OF AIR

The fate of an object hitting the atmosphere depends on its size.

The Earth's atmosphere weighs as much as a layer of water 10 meters thick. To figure out whether a meteor is likely to make it through, you can imagine that it's literally hitting a 10 meter layer of water. If the object weighs more than the water it would have to push aside to reach the surface, it will probably make it through. This works pretty well for a rough approximation!

Very large objects—house-size or larger—have enough inertia to punch through the atmosphere and hit the ground without losing much speed. These are the objects that leave craters in the ground.

Small objects—anything from pebble-size to car-size—are too small to smash through the atmosphere. When they hit it, they heat up until they break apart, evaporate, or both. Sometimes, pieces of these objects survive entry into the atmosphere, either because other pieces absorb the heat and shield them, or because they're made of a material that can withstand the reentry conditions. But when they do, they lose their orbital speed and then fall at terminal velocity straight down to the ground. After the brief pulse of heat during breakup, this free fall through the cold upper atmosphere takes several minutes, which is why meteorites are often very cold when they're found.

These surviving bits of debris hit the ground at relatively low speeds. If they land in soft dirt or mud, they can splash a little, but they don't leave much of a crater. This is why all impact craters on Earth are large: only large, heavy objects keep their orbital kinetic energy all the way to the ground. There are impact "craters" a few feet across—barely larger than the objects that made them—and impact craters a few thousand feet across, but nothing in between.

WILL IT MAKE IT THROUGH THE ATMOSPHERE?

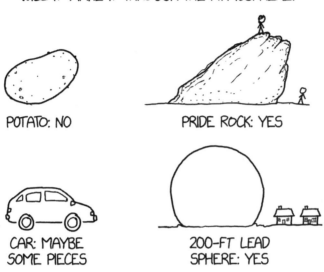

POTATO: NO

PRIDE ROCK: YES

CAR: MAYBE
SOME PIECES

200-FT LEAD
SPHERE: YES

Without shielding, spacecraft break up in the atmosphere. When large spacecraft enter the atmosphere without a heat shield, between 10 percent and 40 percent of their mass usually makes it to the surface, and the rest melts or evaporates. This is why heat shields are so popular.

To protect your package on the way down, you can use a heat shield, too. The easiest kind is an *ablative* heat shield, one which burns away as it goes. It's not reusable, like the heat-resistant tiles on the Space Shuttle, but it's simpler and can handle a wider range of conditions. Then, you just need to shape the capsule so that it points in the right direction—heat shield in front, package in back—and send it on its way.

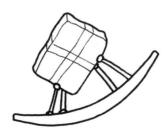

You may also want to add a parachute, for the final drop, but if your package is something lightweight or durable, like socks, paper towels, or a letter, it might be able to survive the final terminal velocity fall relatively undamaged.

Every human-built object which has been designed to survive reentry has used a curved protective heat shield—with a few exceptions.

THE APOLLO SUITCASES

The Apollo program sent seven teams of astronauts to land on the Moon. Each crew carried, among other things, a suitcase-size "experiments package" which would be left on the Moon's surface to take measurements and transmit information to Earth. Six of the seven were powered by radioactivity from plutonium. (The first experiments package, on Apollo 11, was simpler. It had solar power for its electronics, but still used plutonium heaters to keep it warm.)

Six of the Apollo teams landed on the Moon and deployed their suitcases. One of them, Apollo 13, famously did not. After part of their spaceship exploded,[2] they aborted the mission and flew back to Earth. Everyone was ok, it was very heroic, etc. But let's talk about that suitcase.

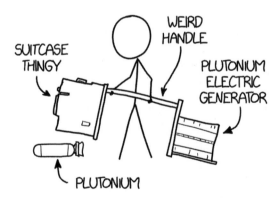

Since the astronauts didn't make it to the Moon, they couldn't leave the plutonium-filled suitcase there, and it came back with them to Earth. That created a problem.

Only the command module, with the astronauts inside, was designed to safely return to the Earth's surface. The other parts of the spacecraft, including the lunar lander, were designed to burn up in the atmosphere. The command module only had enough room for

---

2   It's not as bad as it sounds. Ok, it was roughly as bad as it sounds.

the astronauts and their samples. The suitcase—and the plutonium core, which was stored separately—would have to stay behind in the doomed lander. But if the container holding the plutonium broke apart, it would scatter the radioactive material into the atmosphere.[3]

Luckily, the engineers behind the suitcase had anticipated this possibility. The plutonium was contained within a high-strength cask, about the size and shape of a small fire extinguisher, shielded by layers of graphite, beryllium, and titanium. The protective shell would allow it to survive reentry, even as the rest of the discarded lunar module broke apart violently around it.

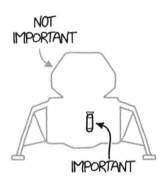

When the Apollo astronauts climbed into the command module as they approached Earth, they left the suitcase behind in the lunar module; then they fired the lunar module's engines to divert it to the area over the Tonga Trench, one of the deepest parts of the Pacific—so the cask would fall into the sea and sink to the bottom. In the decades since, no excess radioactivity was ever detected, which means the protective shell did its job. The cask of plutonium lies on the floor of the Pacific to this day. The plutonium is about half-decayed by now, but it's still producing over 800 watts of heat as of 2019. Maybe some deep-sea critter looking for warmth is cuddled up to it right now.

---

3   On the other hand, this *was* the mid-20th century—you'd think if they were so worried about radioactive particles in the atmosphere, perhaps they should have considered not setting off so many nuclear bombs. But what do I know; I wasn't there.

SEND A LETTER

One of the best ways to get around the engineering challenges of reentry might be to ditch the heat shield entirely in favor of a simpler solution: a manila envelope.

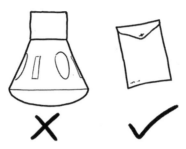

Lightweight objects that experience more drag start slowing down at a higher altitude, where the air density is lower. Since the air is so thin, it doesn't heat the object as efficiently, and although the reentry takes longer, peak temperatures can be much lower. In fact, calculations by Justin Atchison and Mason Peck have shown that an object shaped like a sheet of paper, curved to fall flat side first, could in theory enter the atmosphere "softly" without *ever* reaching especially high temperatures.

If you print your message on a sheet of baking parchment paper, aluminum foil, or some other thin and lightweight material which can survive being warmed up, you might just be able to toss it out the door as is. As long as it's shaped right, it could make it to the ground intact. In fact, a team of Japanese researchers planned to try this by launching paper airplanes from the ISS. They designed the planes to survive the heat and pressure of reentry, but, sadly, the project never went through.

A package tossed by hand from the ISS will descend gradually over the course of many orbits, with little control over the eventual landing point. Controlling where the package will land is much harder than simply delivering it to Earth.

Returning spacecraft generally try to control where they land. Some do this with more precision than others. SpaceX's spent rocket boosters can guide themselves precisely enough to land directly on a target on the deck of a boat, while the older Apollo and Soyuz spacecraft have generally missed their targets by a few miles.[4] Spacecraft undergoing uncontrolled reentry—like your package—can miss their intended landing site by hundreds or thousands of miles.

---

4   The Apollo command modules landed in the ocean. Soyuz spacecraft land in a large open area in Kazakhstan where they're not likely to hit anything.

You can improve the precision of your package delivery by throwing the package really hard. A fast throw can get the package down into the atmosphere more directly, without a long delay as atmospheric drag causes its orbit to slowly decay in a hard-to-predict way. Surprisingly, the way to do this isn't to throw the package downward, toward Earth. Instead, you should throw it backward. If you throw it downward, it will still have enough forward speed to stay in orbit—it will just be a slightly different orbit. You want it to *lose* speed instead.

The faster you throw the package, the more precise its landing. The ISS is traveling at almost 8 kilometers per second, but luckily, you don't need to throw your package that fast. Shaving off just 100 meters per second from the orbital speed at the ISS's altitude is enough to deliver your package to the atmosphere. Unfortunately, throwing something at 100 m/s is difficult. Even the fastest pitchers don't break 50 m/s. Golf balls, on the other hand, travel fast enough. A golfer floating next to the ISS could conceivably hit a golf ball out of orbit in a single stroke. If your package is the size of a golf ball, you can try that delivery method.

If you launch the package at 100 m/s, it will enter the atmosphere traveling at a downward angle of about 1°, which will give you a *debris footprint*—the area where your package might land—over 2,000 miles long. If you're aiming for St. Louis, it could land anywhere between Montana and South Carolina. If you can throw it harder—250 or 300 m/s—you can enter the atmosphere at a proportionally steeper angle and cut the debris footprint down to a few hundred miles. However, no matter how fast and precise your throw, the randomness of turbulence and wind will keep you from hitting a target with a precision of better than a few miles.

MIR

In March of 2001, the space station *Mir* was about to reenter the atmosphere. Most of it was expected to burn up, but some of the larger modules had a chance of making it to the surface. The Russian Mission Control planners tried to time its reentry so it would come down over an uninhabited region of the Pacific, but no one knew exactly where it would land.

Capitalizing on this, Taco Bell came up with a unique promotion: they floated a giant sheet on the Pacific with a bull's-eye painted on it, and offered a free taco to everyone in America if any piece of *Mir* hit the target.

Sadly, no debris ultimately hit the bulls-eye.[5] Most of the larger pieces hit the surface of the ocean in the vicinity of 40°S 160°W – the "spaceship graveyard," a region far from land where the wreckage of over 100 spacecraft has splashed down – and sunk to the bottom.

Despite the claims of many eBay auctions, no confirmed *Mir* debris was ever recovered. If you do find some, you could always try bringing it to Taco Bell headquarters in Irvine, California. Maybe they'll let you exchange it for a taco.

ADDRESSING

You may not be able to aim your package very carefully, but don't despair – that doesn't mean it can't be delivered! You just need to figure out what address to write on it. But, as the US government learned in the 1960s, figuring out what to write on space packages can be tricky.

The first US spy satellites used film cameras. After they had taken their photos, the capsules containing the film were dropped back to Earth. If all went well, they would be tracked on the way down, and an Air Force aircraft would literally catch them out of the air using a long hook.

I CAN'T BELIEVE THIS WORKED.

Things didn't always work as planned. Several capsules returned to Earth uncontrolled; one, which came down in the Arctic near Svalbard, was never found. In early 1964, a Corona reconnaissance satellite – after taking a few hundred photos – broke down in orbit, stopped responding, and headed toward an uncontrolled reentry. Government officials

---

5   Was Taco Bell serious about this? Well, sort of. They bought a $10 million insurance policy to cover the demand for free tacos in the unlikely event of a "win." This policy was purchased from SCA Promotions, a company which provides coverage for promotional contest wins. When a company wants to promise a large prize to anyone who completes a difficult task, they pay a fixed amount to SCA Promotions, and SCA pays out if they succeed. However, the premiums Taco Bell paid for the policy probably weren't too high – since they placed the target near the Australian coast, thousands of miles west of the reentry path.

watched anxiously, trying to determine where it would enter the atmosphere. Eventually, it became clear that it was going to land somewhere in the vicinity of Venezuela.

Observers in the area were told to watch the skies, and on May 26, 1964, debris was seen streaking over the Venezuelan coast.

The officials thought it had probably landed in the ocean, but it had in fact fallen on the border between Venezuela and Colombia. It was found by some farmers, who took it apart, removed the gold discs they found inside,[6] and tried to put the rest of it up for sale. A farmer used the parachute lines to make a harness for his horses. When no one wanted to buy it, the capsule was handed over to Venezuelan authorities, who contacted the United States.

Up until 1964, the returning capsules were labeled UNITED STATES and SECRET in threatening letters, intended to dissuade people from opening them and accessing their highly classified cargo. After the 1964 incident, the United States changed their labeling strategy. Instead of a stern warning, they simply stamped them with a message – in eight languages – promising a reward for bringing the capsule to the nearest US consulate or embassy.

If you want to maximize the chance that the person who finds your package helps deliver it to its intended recipient, bribery might be the way to go.

---

6   The gold discs were part of a science experiment. The science experiment was part of the cover story, in case anyone asked what the satellite was doing up there.

CHAPTER 16

How to Power Your House

(on Earth)

You've got a home full of stuff that needs to be plugged in. How do you power your house?

BONK
BONK

A typical American household consumes about a kilowatt of power averaged over the year. At 2018 electricity rates, that works out to $1,100 per year. Could your land offer a cheaper alternative?

Let's take a look at some of the different sources you might tap into, using a typical US home as an example.

A median newly built single-family home in the United States sits on a 0.2 acre lot (800 square meters), 25 percent of which is occupied by the house itself. Let's assume you live in such a house, and consider what flows of power your little plot of land gives you access to.

Traditionally, when you owned land, you also owned the column of air above your land and the dirt under it—as expressed in the maxim *Cuius est solum, eius est usque ad coelum et ad inferos*, which means, "The property you own extends upward to heaven and downward to hell."

In modern times, your ownership upward may be restricted in various ways—including local zoning laws, the Federal Aviation Administration, and the Outer Space Treaty of 1967, which prohibits claims of ownership over outer space. Your ownership downward may also be limited by the fact that mineral rights are often sold separately from the land – so you may own the property, but not everything buried under it.

But assuming you have complete ownership, here are some of the resources you can expect to find in those three regions:

PART 1: SOLUM

Plants

Plants grow on land, sometimes so enthusiastically that you have to do a lot of work to stop them.

Plants can be burned for fuel, although it's not necessarily the cleanest or most efficient way to produce electricity. If you grow and harvest trees on your land, you could get a steady supply of power by burning the wood.

Timberland productivity depends on management practices, but the National Association of Conservation Districts estimates that the supply of wood from a 3,987-acre pine forest with relatively hands-off management can supply 1 megawatt of ongoing electrical power. That means that if you filled your yard with trees (excluding the 25 percent taken up by your house) then the power you could produce would be...

$$\frac{0.2 \text{ acres} \times 75\% \times 1 \text{ megawatt}}{3,987 \text{ acres}} = 38 \text{ watts}$$

...38 watts. That's enough to charge your phone or run a tablet or small laptop, but not nearly enough to power your whole house.

Other crops might be more efficient—switchgrass, for example, could produce about a kilowatt per acre in much of the central United States, and potentially double or triple that in other areas. Unfortunately, even if you planted it on your roof as well as your yard, that wouldn't be enough to power your house.

Water

Water flows over the ground under the influence of gravity, and this gravitational energy can be harvested by hydroelectric turbines.

The United States averages about 31 inches of rain over its whole land area, and has an average elevation of 2,500 feet. If the country were a uniform plateau 2,500 feet high, with rain falling across it and spilling over the edges...

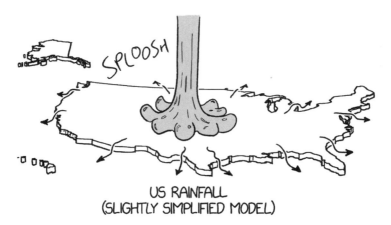

US RAINFALL
(SLIGHTLY SIMPLIFIED MODEL)

...it would generate a total of 1.7 terawatts of power:

$$\frac{31 \text{ inches}}{\text{year}} \times \text{US land area} \times \text{water density} \times 9.8 \tfrac{m}{s^2} \times 2,500 \text{ feet} = 1.7 \text{ TW}$$

The United States has about 120 million households, so that's 14 kilowatts per household!

Unfortunately for your house, that's a very optimistic assessment. Most of the United States' rain falls at lower elevations, and not all of it flows into easily harnessed streams. The Department of Energy suggests the total available United States hydropower—which would include building dams on wildlife reserves and scenic rivers—is 85 gigawatts, 1/20th of that total. That's just 700 watts per household.

PART 2: INFERNO

Buried fuel

If your 0.2-acre property represents 1/12,000,000,000th of the United States, let's imagine that it contains 1/12,000,000,000th of the United States' mineable reserves. Of course, in reality, all those resources are distributed across the country in small pockets, so you either have much more than that, or much less. But if they were evenly distributed, here's what you'd have under your property:

- **3 barrels of oil**. Each barrel of crude oil can provide about 6 gigajoules of power, so 3 barrels is enough to power your house for about 8 months.
- **38,000 cubic feet of natural gas**, enough to power your home for a little over 16 months.
- **19 tons of coal.** Coal has an energy density of roughly 20 megajoules per kilogram, so your 19 tons of coal would be able to power your house for 12 years.
- **1½ ounces of uranium,** which would power your house for a few months in a traditional nuclear reactor, or over a decade in an advanced type of reactor called a "fast neutron reactor." Fast neutron reactors are much more efficient, but they're also much more expensive to operate, and they involve enriching uranium closer to the level where it might be useful for nuclear weapons, so they can make international regulatory agencies nervous.

When you add them all up, those buried fuels represent a few decades worth of power.

In reality, your plot of land wouldn't really contain all those fuel deposits – in all likelihood, it contains *none* of them. And even if it did, the energy it would take for one homeowner to dig them up would outweigh the power they'd generate. Besides, in terms of impact on the Earth's climate, humans certainly can't afford to burn all the fossil fuels hidden in the ground, so maybe it's just as well that you leave them.

Geothermal power

The Earth is still cooling off, both from the heat generated when it initially collapsed into a ball, and from the heat generated by radioactive decay of potassium, uranium, and thorium deep within the planet. The planet cools by radiating heat through its surface. In most places, this heat is usually very faint and hard to detect. In a few places, it's very hard to ignore.

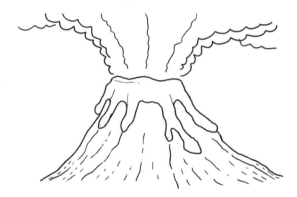

The heat flow in a typical geologically quiet area might be 50 milliwatts per square meter, so in principle your property should have access to 40 watts of heat power indefinitely. Actual geothermal power production involves drilling wells deep into the Earth, pumping water down, and letting the warm rocks heat the water. The heat reservoir would be replenished from the surrounding areas, so you would really be drawing heat from under *everyone*.

In practice, geothermal power is only practical in geologically active areas where high temperatures can be found close to the surface. The Geysers, a sprawling geothermal plant in northern California, produces about 77 kilowatts of power per acre, so if you happen to live there, you could easily power your house. In more geologically quiet areas, geothermal power is likely to be – at best – a source of a little extra hot water.

Tectonic plates

Living on a fault line would have its downsides, but perhaps you could find ways to take advantage of it. The ground exerts a force over a distance, and force times distance is energy. An inch of movement a year isn't much, but that movement has a virtually infinite amount of force behind it. Could you harness it to generate electricity?

In theory, yes!

Suppose you built a pair of giant pistons, anchored to a large area of crust on each side of the fault, and had the pistons compress a reservoir of fluid between them.

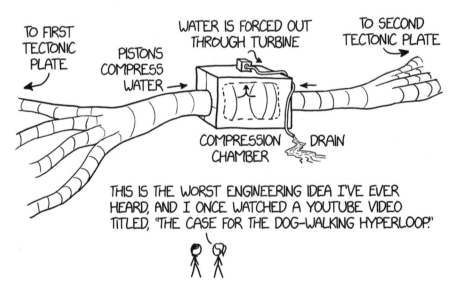

The pressure on the fluid would build up over time, and could be used to drive a turbine. The maximum theoretical pressure this contraption would generate would depend on the pressure the piston could tolerate. If the piston material had a maximum compressive strength of 800 MPa, and the pistons were the width of your yard and twice as high—so the piston heads had a surface area of 0.4 acres—then the total theoretical power available would be given by multiplying the fault movement rate times the area of the piston times the pressure:

$$\frac{1 \text{ inch}}{\text{year}} \times 0.2 \text{ acres} \times 800 \text{ MPa} = 1 \text{ kilowatt}$$

This whole system is ridiculous and technically infeasible for a lot of reasons – and if you tried to build one, you'd probably discover some new ones. But one reason it's a ridiculous idea is cost.

The "roots" of the structure that anchored the generator to the crust would need to extend outward a great distance – otherwise, the crust would simply crack and new fault lines would form. These "roots" would have a volume measured in millions of cubic meters. If they were made of steel and extended 5 kilometers in each direction, they would weigh 60 billion tons and cost something like $40 billion.

Now, $40 billion is a lot of money, but you'd also be saving $1,100 every year on electricity costs. At that rate, you'd make your money back in . . .

$$\frac{\$40 \text{ billion}}{\frac{\$1,100}{\text{year}}} = 36 \text{ million years}$$

. . . 36 million years.

BUT EVERYTHING AFTER THE FIRST 36 MILLION YEARS IS JUST MONEY IN THE BANK!

DON'T YOU HAVE TO KEEP BUYING LAND BECAUSE YOURS IS MIGRATING NORTH?

HEY, YOU GOTTA SPEND MONEY TO MAKE MONEY.

PART 3: COELUM

The Sun

The average sunlight power falling on a plot of land in the United States varies by latitude, cloud cover, and time of year, but a typical value is around 200 watts per square meter. That's an average over the whole year—the power can reach 1,000 watts per square meter when the Sun is highest in the sky, but clouds, seasons, and the fact that it's dark at night bring the average down. (Utilities usually measure things in terms of kilowatt-hours; in those units, 200 watts is equivalent to about 5 kWh per day.[1])

Modern solar panels convert about 15 percent of the Sun's energy into electricity, so if you cover your yard in solar panels, you'll capture 25 kilowatts—much more than you need:

$$0.2 \text{ acres} \times 200 \ \tfrac{\text{watts}}{\text{m}^2} \times 15\% = 25,000 \text{ watts}$$

You could improve your efficiency by tilting the panels to face the Sun, to either cover more area—at the expense of your neighbors—or to get the same amount of power with less ground space...

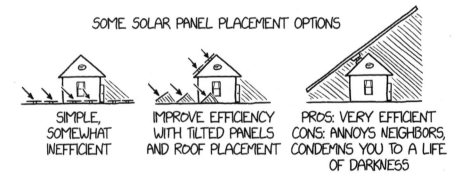

SOME SOLAR PANEL PLACEMENT OPTIONS

SIMPLE, SOMEWHAT INEFFICIENT

IMPROVE EFFICIENCY WITH TILTED PANELS AND ROOF PLACEMENT

PROS: VERY EFFICIENT
CONS: ANNOYS NEIGHBORS, CONDEMNS YOU TO A LIFE OF DARKNESS

... but the effect of this would be relatively small. The limiting factor for solar power generally isn't the available area—it's the cost of the panels. An acre of solar panels might

---

[1] A note on units: "1.38 kilowatts" isn't a per-year measurement—it's just the rate at which an average American consumes electricity, averaged over time. People are used to measuring electricity consumption in kilowatt-hours (the energy to supply one kilowatt for one hour) since that's how it's priced and sold. This is perfectly valid, but it's a little strange from a physics point of view. After all, the average could just be expressed in "kilowatts." It's like saying the width of a road is "100,000 square feet per mile" instead of saying that it's 20 feet wide.

cost over $2 million in 2019 – plus more if you want to be able to store the power in case the Sun disappears.

At 2019 electricity rates of 13 cents/kWh, a solar panel on our example piece of land would pay for itself in 14 years – but various tax incentives, and the opportunities to sell excess power back to the grid, may reduce that "payback period" significantly. In areas with plenty of Sun and/or generous renewable energy incentives, new solar panels can pay for themselves within just a few years.

Wind

The amount of available wind power depends on how windy your area is and how high above your land you're willing to build. In general, wind speeds increase the higher you go, so if you build a taller turbine, you can get more power from it. The US National Renewable Energy Laboratory has mapped out the available wind power potential across the United States for turbines of various heights. The available power is measured in watts per square meter, which can allow you to calculate the power that will pass through a turbine of a given size.

An area like St. Louis, with roughly "normal" windiness, has wind power potential of about 100 W/m² at 50 meters above the ground, 200 W/m² at 100 meters, and perhaps 400 W/m² at 200 meters. Very windy areas, like the Rocky Mountains, might have power densities more than four times that, while in less-windy areas like central Georgia and Alabama, the available power might be a quarter of that.

If your 0.2-acre property is square, then you can fit a wind turbine 28 meters in diameter—or as wide as 40 meters, if the prevailing winds let you put it diagonally.

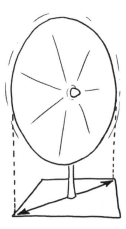

A turbine 28 meters in diameter has an area of 640 m². If it's installed at a height of 50 meters, where the potential is 100 W/m², then the power available will be 64 kilowatts. Wind turbines aren't 100 percent efficient; thanks to Betz's law, they can never extract more than 60 percent of the energy of the wind passing through them. In practice—thanks to varying wind speeds and conversion losses—the actual power captured is closer to 30 percent of the average available. Still, 30 percent of 64 kilowatts is 19 kilowatts—enough to power your house *and* the houses of 18 of your neighbors.

That goodwill might come in handy, since a 28-meter wind turbine 50 meters off the ground might cause some problems on your street. The bottom of the blade will be just 36 meters above the ground, so hopefully you don't have any unusually tall trees.[2] And you should probably discourage neighborhood kids from flying kites.

---

2   If you do, you won't for long.

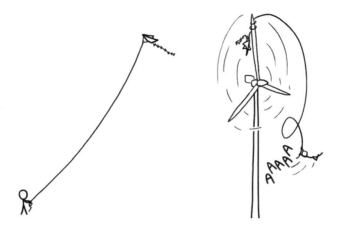

EPILOGUE: SPACE ITSELF

Some theoretical models of the universe suggest that the quantum fields that make up space exist in what's called a "false vacuum." After the Big Bang, the fabric of space settled from a high-energy chaotic quantum froth into its current form. Under these models, the form it settled into isn't really settled – space-time itself contains a certain amount of tension, and if perturbed in the right way, this tension could be released, and space would fall into a fully relaxed, settled state.

In these models, the false vacuum represents a huge amount of potential energy in every cubic meter of space. Your yard has lots of easily reachable space – could you trigger vacuum decay and solve your problems forever?

To answer this question, I contacted astrophysicist and end-of-the-universe expert Dr. Katie Mack. I asked Dr. Mack how much power would be released if someone triggered vacuum decay in their yard, and whether it could be harnessed to power their home. Her response: "Please do not do that."

"If you could decay the vacuum locally, it would in principle release the energy of the Higgs field, probably in the form of extremely high energy radiation," she said, "but along with that energy you'd get a bubble of true vacuum expanding at the speed of light, making it impossible to harness any of the energy before the bubble envelops you. That true vacuum bubble would incinerate you, then destroy all your particles, and then devour the entire Universe. And then immediately collapse it."

 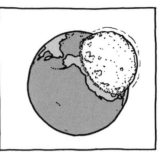

Luckily for us, the fact that the universe has existed for this long without decaying suggests that, even if the false vacuum theories are correct, vacuum decay isn't particularly *likely* any time soon.

"Even though vacuum decay is pretty much inevitable if our current understanding of particle physics is correct, it's astronomically unlikely to happen any time in the next several trillion years. There are better and more efficient ways to get energy," added Dr. Mack. "For instance, why not create a tiny black hole and use the radiation from its Hawking evaporation like a campfire? Depending on the mass, you could get a nice steady glow that lasts for years before it spectacularly explodes in its final disintegration!"

That sounds much more practical.

CHAPTER 17

How to Power Your House

(on Mars)

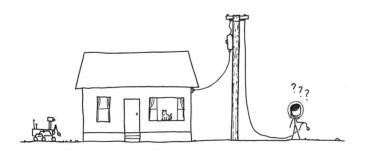

Power is harder to come by on Mars than on Earth.

This is partly for the obvious reason that there's no electrical grid. But even if we build one there, the usual sources we draw on for electrical power on Earth won't work as well on Mars.

Type of power	Does it work on Mars?	Reason
Wind power	Not really	The air is too thin
Solar power	Not as well	The Sun is farther away
Fossil fuels	No	No fossils
Geothermal power	Not as well	Not very much geologic activity
Hydropower	No	No rivers
Nuclear power	Not unless you bring your own fuel	Certain geologic processes needed to concentrate uranium
Fusion	No	Doesn't even work on *Earth*

But there's one very unusual potential source of power on Mars. You just need to be willing to destroy a moon to get it.

You don't have to feel bad about destroying Mars's moon Phobos – it's already doomed.

Earth's moon orbits more slowly than Earth spins, so the tidal drag between Earth and the Moon slows down the Earth and speeds up the Moon. Since the drag speeds up the Moon, it flings it progressively farther away.[1] On Mars, the situation is different: Phobos orbits *faster* than Mars spins, so tidal drag pulls it backward, causing it to fall down into a tighter orbit. Over time, Phobos is moving closer and closer to Mars.

---

1   For more on this, see chapter 27: How to Be on Time.

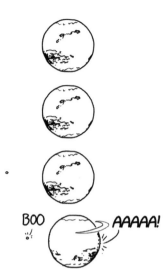

Phobos may not be very heavy compared to other moons – Earth's moon is 7 million times more massive – but it's still pretty big by human standards.

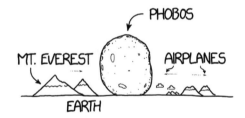

Phobos's mass and speed mean that it carries a huge amount of kinetic energy as it circles Mars – energy you can potentially tap into.

PHOBOS TETHER

Attaching a tether to Phobos has been proposed before. Normally, the goal of a Phobos tether proposal is to use Phobos's position and orbital energy to efficiently move large amounts of cargo to and from the surface of Mars, which often involves using one end of the tether as a "skyhook" to grab cargo leaving Mars's surface.

But a tether could also be used to extract energy from Phobos directly. If you attached a 5,820-kilometer tether to the Mars-facing side of Phobos, the end of the tether would dangle into Mars's atmosphere. The dangling end would be moving through Mars's atmosphere at 530 meters per second (m/s). On Earth, that would be about 1.5 times the speed of sound, but because Mars's atmosphere is mostly carbon dioxide, sound travels more slowly there,[2] and 530 m/s is 2.3 times the Martian speed of sound.

WIND TURBINES

Wind turbines on the surface of Mars aren't all that useful because the air is so thin and slow-moving that it would have trouble even turning a turbine blade. But the end of the tether would experience wind blowing past at Mach 2.3 – and *that's* a different story. The air flowing past the tether would carry around 150 kilowatts of energy per square meter. A turbine 20 meters in diameter would potentially have 50 megawatts of energy passing through it, enough to power an entire town.

2   Since the speed of sound on Mars is lower, if you tried to talk, your voice would sound significantly deeper.

Wind turbines aren't normally designed to work at supersonic speeds, because supersonic winds are rare on Earth outside of meteor impacts, volcanic blasts, and nuclear shock waves. But there *are* some turbines designed to attach to supersonic aircraft or rockets. These turbines are designed to generate power from the flow of air past the fuselage, in part to power the aircraft's systems if the engines die. The supersonic turbines are streamlined and have short, stubby blades; your Mars turbine will likely need to look more like those than the typical wind turbine.

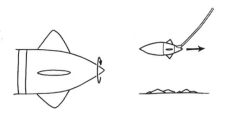

Your turbine will be dragged through Mars's atmosphere by Phobos, which will sap the moon's momentum and cause it to spiral inward. The more turbines you add, the more power you'll generate and the faster Phobos will descend. Note: As Phobos gets closer, you'll need to shorten the tether to keep it from hitting the ground. Helpfully, the shorter tether won't need to be as massive to support its own weight, so over time you'll be able to support more turbines with the same amount of tether material.

The total energy available by bringing Phobos down to the top of Mars's atmosphere is:

$$G \times \text{mass of Mars} \times \text{mass of Phobos} \times \tfrac{1}{2} \times \left(\frac{1}{\text{Mars radius} + 100 \text{ km}} - \frac{1}{9,376 \text{ km}} \right) \cong 4 \times 10^{22} \text{ J}$$

Each American uses, on average, 1.38 kilowatts of electricity, which means Phobos's orbit carries enough energy to supply the electricity needs of an America-size population for almost three millennia. Even if lots of neighbors move in, there's plenty of Phobos power to go around.

Space tether projects involve massive amounts of material, and this one will be no different. Even a small tether from Phobos to Mars will weigh thousands of tons, and the weight will increase as you add more and larger turbines. The amount of power produced by a tether turbine is proportional to how much force the tether exerts on it, so each additional watt of turbine capacity increases the strain on the tether, and the tether has to be made more massive to support it. Conversely, we can think of each added kilogram of tether material as "producing" a certain amount of power.

The weight of the tether, and the efficiency, will depend on the materials you use and a lot of engineering details, but overall the tether will probably deliver, at most, 2 watts of power for every kilogram of tether. Since the tether can produce that energy indefinitely, over a timescale of decades, that 2 watts per kilogram adds up to far more total energy per kilogram than common fuels like batteries, oil, or coal.[3]

The turbines will be inefficient to a degree that's hard to predict. Since the airflow is effectively unlimited, your main concern will be reducing the "wasted" drag on the tether, rather than capturing all the power of the air that passes through it. It's possible that alternate turbine designs will prove more efficient and reliable—you may want to experiment with designs such as Darrieus turbines, drag-based turbines, or Magnus effect turbines, all of which find specialty uses here on Earth:

WIND TURBINES

REGULAR   DARRIEUS   DRAG   MAGNUS

In addition to inefficiencies associated with the turbines, you'll need to worry about getting the power from the turbine to your house on the surface, which will inevitably involve additional losses. Power transmission could involve anything from beamed microwave power to dropping huge numbers of rechargeable batteries down to the surface.

PHOBOS   MARS

When a moon orbits too close to a parent body, tidal stresses can become strong enough to pull material from the moon's surface. The distance at which this happens is called the Roche limit. As Phobos draws closer to Mars, it may break apart into a debris ring. To keep

---

3 It doesn't come close to plutonium, though, which produces hundreds of watts of heat per kilogram for many decades. But plutonium is hard to come by in large amounts. The *Curiosity* rover—possibly your Martian neighbor—is powered by a 5-kilogram chunk of plutonium, acquired by NASA at great expense.

this from happening, you may need to use some kind of high-strength net to hold Phobos together, or let it break up into several smaller moons, each of which can be held together with netting more easily.

This kind of orbital turbine has a particularly strange feature: the longer you use it, the more power it provides. Your tether will exert drag on Phobos, which causes the moon to descend . . . but as it descends, it also speeds up, because lower orbits are faster. A faster orbit means a faster-moving tether, which means faster airflow and more turbine power. The tether will produce steadily more power over the lifetime of Phobos.

WHEN PHOBOS LANDS

Eventually, once drag has withdrawn the full 4×10^{22} joules of energy from Phobos—perhaps millennia in the future, or perhaps in just a few years, depending on how much power your house uses and whether other colonists are tapping into the turbines as well—the moon will reach Mars's atmosphere.

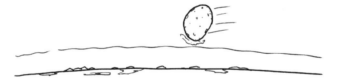

Phobos is similar in size to the rock that collided with Earth at the end of the Cretaceous—a collision which led to the extinction of most of the dinosaurs. Phobos's impact with Mars, whether it is in one piece or several at that point, will be similarly disruptive. The tether, over thousands of years, will have consumed gravitational potential energy from Phobos and delivered a total of 4×10^{22} joules of it to the planet, while at the same time causing Phobos to accelerate as it descends. The impact of Phobos with the surface will deliver a similar amount of energy, but all at once.

Phobos's impact will leave a long scar girdling Mars, and the collision will spray a huge volume of debris into space, most of which will fall back down in a rain of molten rock reaching every part of the surface. As is so often the case, a "free" source of energy ultimately comes with a terrible long-term cost.

The apocalyptic consequences won't be *all* negative. For a brief time, until the lava rain dies down, some of Mars's lower valleys may heat up enough so that liquid water could exist in stable pools on the surface.

If your house happens to be located in one of these valleys, see chapter 2: How to Throw a Pool Party.

CHAPTER 18

How to Make Friends

If you just start walking, eventually you'll bump into someone.

This might take a while. You might be lucky and walk right into a crowd of people, but if you're in a sparsely inhabited area, it could take weeks. If you start walking from a random location in an area containing some number of people, you can calculate the time it will take to run into someone by using the physics concept of a *mean free path*:

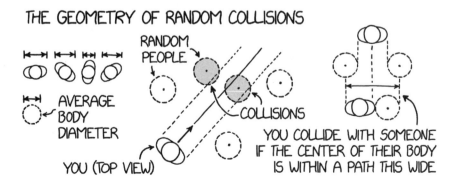

$$\text{time per collision} = \frac{1}{\text{collisions per hour}} = \frac{1}{(\text{shoulder width} + \text{average torso diameter}) \times \text{speed} \times \text{population density of area}}$$

Some areas certainly make encounters easier than others. Here's the average collision interval for a few different regions:

- **Canada:** 2.5 days
- **France:** 2 hours
- **Delhi:** 75 seconds
- **Paris:** 40 seconds
- **Mercedes-Benz Stadium in Atlanta during a sold-out game:** 0.6 seconds
- **The field during the game:** 3 minutes

It's clear that if you want to physically run into people, you'll have better luck in a packed football stadium than in the boreal forests of Canada. And if you *do* try the stadium, you'll have more collisions in the stands than on the field — although the collisions on the field will probably be more jarring.

FRIEND!!

But most of the time, random encounters don't lead to friendships. This is ok. Occasionally, you may hear someone complain that people walking around in public need to be shaken up from their routines, that they're too wrapped up in their own little worlds. But people have their own lives. They're not necessarily looking for a connection at the moment you are.

So if it's so difficult to connect, how do people ever make friends at all?

We can get some insight into where people meet their friends through surveys. A Gallup survey of Americans in 1990 asked people where they met most of their friends. The most popular answer was work, followed by school, church, neighborhoods, clubs and organizations, and "through other friends."

A more comprehensive survey by Dr. Reuben J. Thomas, published in *Sociological Perspectives*, asked 1,000 US respondents for information about how they met their two closest friends. The study used the answers to build up a profile of how friendships form at different ages.

Some sources of friendships remained relatively steady—at all ages, people made about 20 percent of their new friendships through family, mutual friends, religious organizations, or encounters in public settings. Other sources of friendships wax and wane throughout life—at first school dominates, followed by work. Then, as people approach retirement age, they become more likely to make friends around the neighborhood and at volunteer organizations.

ADAPTED FROM THOMAS, REUBEN J. 2019. "SOURCES OF FRIENDSHIP AND STRUCTURALLY INDUCED HOMOPHILY ACROSS THE LIFE COURSE." *SOCIOLOGICAL PERSPECTIVES*. DOI: 10.1177/0731121419828399

If nothing else, these studies help answer the question of where people make friends. Those aren't necessarily the places that you should go to maximize your chances of making new friends, but they're where most friendships begin.

Once you do encounter someone, how do you turn the acquaintanceship into a friendship?

GO ON, INTO THE HOPPER.

ACQUAINTANCES

FRIENDS

Here's the bad news: there's no magic formula or trick that can make someone your friend. If there were, that would mean you could apply it to someone regardless of who they were or how they felt. And if you don't care who someone is or how they feel, then you're not their friend.

Immanuel Kant developed a rule called the "categorical imperative," which was at the center of his idea of ethics. He expressed the rule in several different formulations; the second formulation read, in part, *"Act in such a way that you treat humanity . . . never merely as a means to an end, but always at the same time as an end."*

In Terry Pratchett's novel *Carpe Jugulum*, the character Granny Weatherwax expressed the principle more succinctly. A young man tried to tell Granny that the nature of sin was a complicated thing. She said that, no, it was very simple. *"Sin,"* she said, *"is when you treat people as things."*

Whether or not you buy into the philosophy of the categorical imperative, it's good practical advice, because people can tell when they're being treated as things. Whatever our faults, humans have countless millennia of experience in judging the intentions of others—a skill much older and deeper than our ability to put our feelings into words. We can be shortsighted and confused and make lots of mistakes, but we can smell disdain and condescension from a mile away.

So while *encountering* people might be easy, there's no single set of steps you can follow to *befriend* them—because friendship means caring about how people feel. And there's no way to decide how they feel yourself, regardless of how much research or thinking you do. You just have to ask them . . .

. . . and listen to what they have to say.

How to
Blow out
Birthday Candles

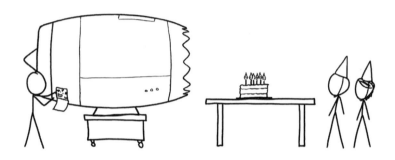

How to
Walk a Dog

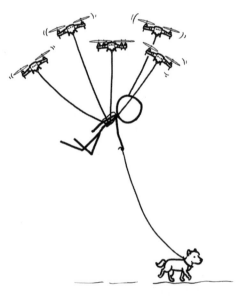

SKREEEEE!

How to Send a File

Sending large data files can be difficult.

Modern software systems have moved away from the concept of "files." They don't show you a folder full of image files; they show you a collection of photos. But files linger on, and will probably continue to do so for decades to come. And as long as we have files, we'll need to send them to people.

COMPUTER
CONTAINING FILE

PERSON YOU WANT
TO SEND FILE TO

The simplest, most obvious way to send a file is to pick up the device the file is stored on, walk over to the intended recipient, and hand it to them.

Carrying computers can be difficult—especially the earlier ones that were the size of a whole room—so rather than carry the whole computer, you can try detaching a piece of the computer containing the file. You can then bring this piece to the other person and let them transfer it to their own device. On a desktop-style computer, the files may be stored on a hard drive, which can often be removed without destroying the computer.

On some devices, though, file storage is permanently attached to the electronics, making removal more challenging.

A more convenient and less destructive solution is removable storage. You can make a copy of the file, put it on a device, then give the device to the person.

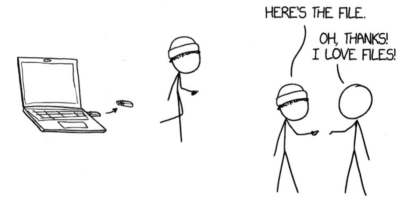

Carrying storage devices around is a surprisingly high-bandwidth way to transfer information. A suitcase full of MicroSD cards contains many petabytes of data; if you want to transfer very large amounts of data, mailing boxes of disk drives will almost always be faster than transferring them over the internet.[1]

If you want to send data to a specific location that's too far to walk, but not convenient to reach by mail—say, a nearby mountaintop—you could try using some kind of autonomous vehicle to carry it. A delivery drone, for example, could easily carry a small satchel of SD cards containing terabytes of data.

---

1   For more on this, see "FedEx Bandwidth" in *What If?*

Quadcopter-style drones don't work very well over long distances thanks to the limitations of batteries. If a drone has to carry its own battery, it can only hover for so long. If it wants to hover longer, it needs to carry a bigger battery, but that means more weight and faster power consumption. For the same reason that a house supported by jet engines can only hover for a few hours,[2] small coaster-size drones typically have flight times measured in minutes, and the larger ones used for photography are usually limited to less than an hour in the air. Even if it flew very fast, a tiny drone carrying a MicroSD card could make it just a few miles before running out of steam.

You could increase your range by making the drone bigger, adding solar panels, flying higher, and going faster. Or you could turn to the real masters of efficient long-distance flight:

Butterflies.

---

2   See chapter 7: How to Move.

Monarch butterflies travel thousands of miles during their migration across North America, with some traveling all the way from Canada to Mexico in a single season. If you look up during the spring or fall on the East Coast of the United States, you can sometimes spot them gliding by silently overhead, a few hundred feet above the ground. Their extreme range puts drones—and even many large aircraft—to shame.

You might think butterflies have an unfair advantage over battery-powered aerial vehicles, since they can stop to consume nectar and "recharge." Butterflies will certainly refuel if they can, but they don't necessarily need to. Another butterfly species, the painted lady (*Vanessa cardui*), is even more impressive: it flies from Europe to central Africa, a 4,000-kilometer flight that takes it over the Mediterranean Sea *and* the Sahara desert.

Butterflies make these journeys powered only by small reserves of stored lipids. They can fly so much more efficiently than drones in part by soaring – they seek out thermal columns and mountain waves, then hold their wings steady and ride the rising air upward like a vulture, hawk, or eagle.

If you want to send your file to someone who lives along the migration route, could you get a butterfly to carry it for you?

Butterflies can carry weights. Volunteers with groups like Monarch Watch tag tens of thousands to hundreds of thousands of monarch butterflies each year to track their migration and monitor their population (which has been in decline in recent decades). The smaller tags weigh about a milligram, but monarchs have completed their migration with larger tags that weigh 10 mg or more.

MicroSD cards weigh several hundred milligrams – comparable to the weight of a butterfly – so butterflies would have a hard time carrying them. But there's no reason a storage device can't be made smaller. MicroSD cards contain memory chips, and the storage density of these chips might be up to a gigabyte per square millimeter. Given those sizes, a butterfly could easily carry a tiny chip with a gigabyte of data. If your file is larger than that, you could break it up across multiple butterflies, and send multiple copies for redundancy.

When your data finally arrived at its destination, the recipient would have to check a lot of butterflies to assemble all the pieces of the file. You may need to develop some kind of touchless butterfly scanner that allows them to scan many butterflies at once.

You could avoid that problem – and increase your bandwidth dramatically – by using DNA-based storage. Researchers have stored data by encoding it into a DNA sample, then sequencing the DNA to recover it. Systems like this can achieve densities far beyond anything we do with chips – it's possible to store and recover hundreds of petabytes of data using a single gram of DNA.

Each year, tens to hundreds of millions of monarch butterflies arrive in Mexico to spend the winter together in giant colonies in the mountains. If you tagged ten million of these butterflies with tiny pouches containing 5 mg of DNA storage each, the total capacity of the butterfly armada would be about 10 zettabytes – 10,000,000,000,000,000,000,000 bytes. That's roughly the total amount of digital data in *existence* in the late 2010s.

If the Sun is warm, the winds are favorable, and it's the right time of year, you could use butterflies to send someone the entire internet.

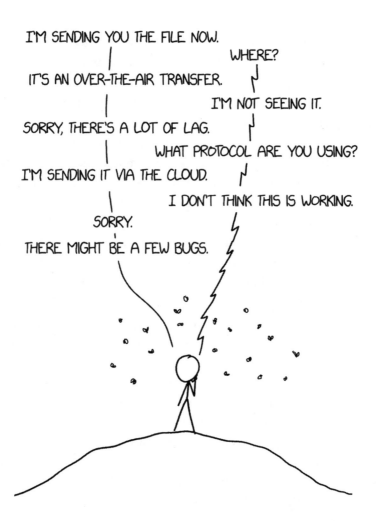

CHAPTER 20

How to Charge Your Phone

(when you can't find an outlet)

The easiest way to charge your phone is to plug it into an outlet. Unfortunately, they're not always easy to find when you need them.

Sometimes, when you find an outlet, there's something already plugged in, such as someone else's phone or an unattended piece of equipment. If you carry a little portable power strip around, you can sometimes unplug the cord for a moment and plug it into your power strip, then use one of the other outlets—although you may want to exercise caution while doing this.

If you can't find an outlet *at all*, your task becomes a little more difficult. Instead of being given energy by a friendly wall, you'll have to take it from the environment some other way.

Humans extract energy from various natural processes. We burn things for heat, we collect energy from sunlight, we take advantage of underground heat, and we harness the movement of wind and water by making them turn the blades of turbines.[1]

In theory, all of these techniques can work indoors as well, but they're a little more difficult. Sure, you can find light, heat, flowing water, and flammable stuff in an airport, but usually in much smaller quantities than outside. This is partly because in an artificial environment, everything is put there by someone. In physics, *energy* and *work* are synonymous. If some human-built contraption is spewing so much energy into the environment that it's worth your time to harvest it, then whoever keeps it running is doing a lot of work for nothing.

---

1   For more on harnessing sources of power outdoors, see chapter 16: How to Power Your House (on Earth).

Unlike most humans, planets and stars have no problem doing work for free.[2] The Sun floods the whole Solar System with light, even the empty parts, and will continue to do so for billions of years without pause – all you have to do is put up a solar panel and capture a tiny amount of it. Indoors, there's less of this energy for the taking, so it's not as easy, but it's still possible. Here are some ways to capture energy in an airport or a mall:

WATER

There may not be actual rivers in an airport, but there's often running water. Water flows out of faucets and fountains, and there's no reason you can't use this water to generate electricity just like a hydroelectric dam does.

You don't need to build a whole tiny hydroelectric dam.[3] Since the building's water system holds the water in a reservoir and directs it into pipes for you, you can skip all that and mount a turbine directly on the mouth of the faucet or water fountain. There are actually companies that manufacture these turbines, either to run small pieces of equipment attached to pipes, or simply as a replacement for a pressure relief valve, to extract some usable energy from water. In the late 19th and early 20th century, many buildings had running water but no electricity, and these sorts of generators – called "water motors" or "hydro-electric dynamos" – enjoyed brief popularity.

The amount of power available from a pipe can be surprisingly large. Moving water carries a lot of energy, and turbines can be very efficient – small turbines can convert 80 percent of the water's energy into electricity, and large ones can achieve even higher efficiencies. A water supply with a pressure of 30 PSI (pounds of force per square inch) and a flow rate of 4 gallons per minute can produce over 40 watts of power, which is enough

---

2   Although there are rumors Jupiter is considering setting up a paywall.

3   Although if you really want to, absolutely, go for it.

to power several LED bulbs, charge dozens of phones, or even run a small laptop with multiple browser tabs open.

In the end, the power you're using is supplied by the pumps run by the water company, which create the water pressure in the first place. Eventually, someone from the airport—or the local water utility—will notice. And even if they don't, 4 gallons per minute adds up quickly. Whether you're paying for the water or not, you have to find somewhere to put it.

Of course, those ramps that connect to the jet slope downward...

WHY IS THE PLANE FILLING WITH WATER?

AIR

Unfortunately, wind power isn't a great choice for capturing energy indoors. There's plenty of air circulating in airports, but "wind" flowing out of a ventilation duct generally carries a lot less energy than water flowing out of a faucet and is harder to capture efficiently. A tiny windmill the size of a handheld portable fan, placed at the exhaust grate of an air conditioning system, could probably produce about 50 milliwatts of electricity—not even enough to keep a single phone charged. Even if you covered an entire exhaust vent with fans, you'd struggle to get even a fraction of the power you could get from a faucet.

This is true outdoors, too—it's easier to get power from flowing water than flowing air. The reason we use air at all is because there's more of it. There's a reasonable chance you're feeling a breeze right now as you read this, but the odds that you're standing in a river are slim. The world has more wind than rivers; the total energy carried by rivers is on the order of a terawatt, while the total energy carried by wind is closer to a petawatt.

FIRE

ESCALATORS

Escalators give energy to their riders. When you get on an escalator and start moving upward, the escalator has to consume extra electrical energy to turn the motors that lift you. This energy is transferred to you in the form of potential energy. If you turn around and slide down the railing back to the lower level, you'll arrive at high speed—having turned the potential energy from the escalator motors, which you got for free from the escalator, into kinetic energy.

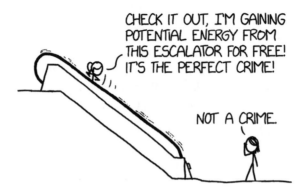

CHECK IT OUT, I'M GAINING POTENTIAL ENERGY FROM THIS ESCALATOR FOR FREE! IT'S THE PERFECT CRIME!

NOT A CRIME.

Escalators may be designed to give you potential energy, but with the help of some simple mechanisms, you can make the escalator give you *electrical* energy instead. Really, escalators are just big metal waterfalls, and you can use the moving stairs to turn an axle, just like a waterfall turning a waterwheel at a mill.

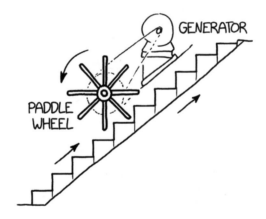

A simple wheel with flat paddles will interlock awkwardly with the escalator. You can make it work more smoothly by building a wheel with curved paddles that mesh with the escalator. If you shape the paddles carefully, the wheel can stay in constant contact with the escalator without sliding.[4]

---

4   The shape of a wheel that can roll smoothly on stairs was derived by Dr. Anna Romanov and classmate David Allen while they were mathematics students at Colorado State University. Their design would work on stairs with a 45° slope and step size; the exact shape of the petals could be tweaked to fit a specific set of stairs.

The amount of power you can extract from an escalator this way can be substantial. The mechanical work an escalator does each minute is simple to calculate—it's equal to the peak number of passengers per minute times weight per passenger times the escalator's height times the acceleration of gravity. When fully loaded with people, a two-story escalator might easily output 10 kilowatts of mechanical power, much of which you could capture with a well-designed wheel. That's not just enough to charge a phone, it's enough to run an entire house.

Pro tip: You should probably make the wheel narrow, rather than having it take up the whole width of the escalator. It's going to be unsafe either way, but if it takes up the whole escalator, and someone gets on without noticing, it will inadvertently turn your contraption into a nightmarish human grinder, which will likely harm its efficiency.

You should use the "up" escalator rather than the "down" one to turn a paddle wheel. It's possible both will work, but the "up" escalator is the one designed to exert *more* force when weighed down by people. A "down" escalator has to do less work when people get on it, since it gets help from gravity, and it may have problems exerting the extra downhill force needed to turn a wheel. You also may want to use multiple wheels, which will help spread out the weight on the escalator.

An escalator waterwheel could extract significant amounts of energy, but that also means it would cost the escalator owners significant amounts of money. If you hook up to an escalator and force it to exert an extra 10 kilowatts of power for 12 hours a day, that could cost the building owners over $400 a month in electricity bills. Needless to say, they probably won't be thrilled if they find out.

If you *are* kicked out of the airport, you should try to take the wheels with you. In addition to working as escalator waterwheels, they can roll down stairs without bouncing, which is a pretty cool property.

Comedian Mitch Hedberg once commented that an escalator can never break; it can only become stairs. Well, an escalator waterwheel generator can never break...

... it can only become an extremely impractical bicycle.

CHAPTER 21

How to Take a Selfie

We sometimes think of our eyes as a pair of cameras, but human visual systems are so much more sophisticated than any camera—it's just easy to miss the complexity because it happens automatically. We look at a scene, get a picture in our heads, and we don't realize how much processing, analysis, and interaction happens to produce that picture.

Cameras generally see all the areas of an image at roughly the same resolution. If you take a picture of this page with a phone camera, a word in the center of the picture will be made up of about the same number of pixels as a word near the edge. But your eyes don't work that way—they see very different amounts of detail at the center of your vision compared to the edges. The actual "pixel grid" of the eye looks very strange:

A CAMERA'S PIXEL GRID   YOUR EYES' "PIXEL GRID"

The reason we don't notice the wildly varying resolutions is that our brains are used to it. Our visual systems process the image and give us an overall impression that what we're looking at is simply what the scene looks like, the same thing that would be seen by a camera. This works . . . until we start comparing our mental picture to what's produced by actual cameras, and discover that there are a lot of variables that our brains have been adjusting for us behind the scenes.

One of the ways in which cameras and eyes can differ is their *field of view*. Field of view is responsible for a lot of confusion in photography, and it has some particularly significant effects on selfies.

When you hold a camera close to your face, it makes your features look different. To understand why—and how it affects photos of all kinds—let's talk about **supermoons**.

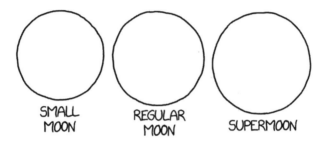

SMALL MOON   REGULAR MOON   SUPERMOON

Every now and then, viral internet stories make the rounds spreading wild claims about some upcoming astronomical event.

 NEXT WEEK, THE MOON WILL COME SO CLOSE TO EARTH THAT YOU'LL BE ABLE TO TOUCH IT FROM SKYSCRAPERS

 ON APRIL 15, A GIANT ASTEROID WILL HIT THE EARTH! SCIENTISTS SAY IT COULD WIPE OUT THE DINOSAURS!!!

 THIS FRIDAY, ASTRONOMERS SAY THE SUN WILL PASS BETWEEN THE EARTH AND THE MOON!

 ON MARCH 24, MARS WILL APPEAR AS LARGE AS EARTH IN THE SKY! SHARE IF YOU AGREE!

 NASA HAS ANNOUNCED THAT NEXT JULY 30, THE ANDROMEDA GALAXY WILL COLLIDE WITH THE MILKY WAY, SO BE SURE TO BRING YOUR PETS INDOORS AND COVER YOUR HOUSEPLANTS TO PREVENT LEAF DAMAGE.

 ON OCTOBER 4, THE SUN WILL BE TURNED OFF FOR 12 HOURS FOR CLEANING.

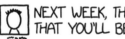 THE ANNUAL PERSEIDS METEOR SHOWER WILL OCCUR AROUND AUGUST 11-12, ACCORDING TO A RADIO MESSAGE ASTRONOMERS RECEIVED FROM *ALIENS!*

These are sometimes accompanied by photos of the "supermoon" behind a skyline, like this.

However, when people go outside to take pictures of the Moon, this is what they get:

So what happened? Was the first photo fake?

It might have been, but often it's not. Instead, it's a photo taken with a very narrow angle, through a telephoto lens.

Every photo shows a certain field of view. A wide field of view shows stuff off to the sides, and a narrow one shows only the objects directly in front of the lens.

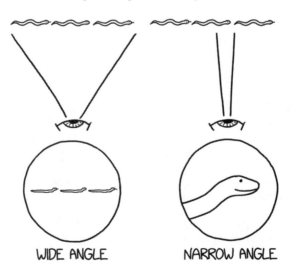

WIDE ANGLE   NARROW ANGLE

"Zooming in" means narrowing the field of view. It's easy to think of zooming in as "getting closer" to the subject, because it makes a small subject get bigger and fill the frame. But zooming in isn't quite the same as getting closer. When you get closer to a subject, the subject gets bigger within the picture, but the distant background stays the same size. When you zoom in, the subject *and* the background get bigger.

ORIGINAL   ZOOMING IN   GETTING CLOSER

The reason people get tricked by this difference is that our eyes have only one field of view. We can focus our attention on something at the center of our vision, but the total area covered by our eyes stays the same. Photos with unusually wide or narrow fields of view can surprise us.

For decades, the rule of thumb among photographers has been that a 50mm full-frame lens produces an image that looks "natural" to people—not too wide and not too narrow. This "natural" lens produces a surprisingly narrow field of view; it's about 40° wide, similar to the area covered by a hardback book when you hold it about a foot from your face.

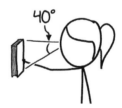

But smartphones may be in the process of changing all that, because phone cameras have *much* wider fields of view than old 50mm lenses.

The iPhone X, for example, has a 65° horizontal field of view, letting users fit a wider scene into the frame without having to back up. (It is not, however, quite wide enough for one common photography subject: rainbows. A rainbow covers 83° of the sky, making it slightly too wide to fit in an iPhone frame.)

These wider-angle lenses may have become more common because smartphone users want to take natural-looking pictures of scenes from life, or selfies that show multiple people. It's hard to take a selfie with a traditional 50mm camera held at arm's length. And phones make it easy to crop our images after the fact, so it makes sense to err on the side of "too wide" and let users do the zooming and cropping. But the wide field of view comes with a cost: when you use a wide-angle lens to take a picture of a small or faraway subject, it may not show you what you expect.

To a human, the Moon is attention-grabbing. Even if we don't literally "zoom in" with our eyes, we narrow our attention to isolate it. We use our high-resolution vision to pick out the details of the Moon, ignoring the comparatively boring sky around it.

But a smartphone doesn't know to "narrow its focus" the way our brain does. The Moon is just another patch of pixels, lost in its extra wide-angle camera. To get a good photo of the Moon, you need to zoom in—something that smartphones have limited ability to do.

HOW THE MOON   HOW THE MOON LOOKS
LOOKS TO MY EYES   TO MY CAMERA

If you do have a camera that can zoom in, all the other stuff you might want to include in the picture—like buildings and trees around you—doesn't fit in the frame anymore. Those things look bigger than the Moon from where you're standing, even though they're obviously not (unless your city has unusually lax zoning regulations.)

DID YOU KNOW
THAT TOWER
IS TEN TIMES
LARGER THAN
THE MOON?

If you want an object to appear small relative to the Moon, you have to move far enough away that it takes up a smaller angle of the sky. For a building, this distance can be pretty large. In order to take one of those photos that shows a huge moon behind a city skyline, the photographer generally needs to stand miles away from the city. That nice-looking photo probably took a huge amount of work and planning.

I HAD TO CLIMB A HILL IN NEW JERSEY AND SIT ON A RIDGE IN THE FREEZING COLD FOR AN HOUR FIDDLING WITH LENSES TO GET THIS PICTURE, SO YOU ALL HAD *BETTER* CLICK LIKE.

The reason buildings look so big in ordinary photos, and the Moon looks so small, is because buildings are so much *closer* than the Moon. And this brings us back to selfies.

WIDE-ANGLE SELFIES

This same wide-angle effect that makes the Moon seem tiny can affect how selfies turn out. When someone takes a photo of their face with a smartphone, their instinct for composition might tell them to hold the phone close enough that their face fills a significant portion of the frame. But at that distance, which is much closer than where someone would usually stand when looking at you, the wide-angle smartphone lens creates an unnatural perspective. Your nose and cheeks are substantially closer to the camera than your ears and the rest of your head, which makes them look bigger—just like a building in the foreground of a smartphone shot looks bigger than the Moon.

This distortion can make faces look subtly different in ways that we don't expect. To reduce the effect, hold the phone farther away and zoom in—either within the camera app as you take the picture, or after the fact by cropping it.

How far away should you hold the phone? To minimize perspective distortion between several objects in a frame, the distance to the phone should be much larger than the difference in distance between the nearest and farthest objects.

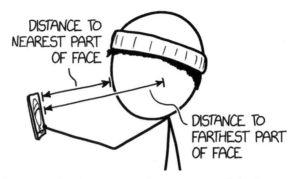

The difference between the distance to the nearest and farthest visible parts of your face is probably less than a foot, which means that the distortion can change a lot depending on whether you hold the camera up at a normal distance from your face or stretch your arm out all the way. Holding a camera 5 or 6 feet away will almost entirely eliminate this kind of distortion, but our arms aren't long enough for that—which partly helps to explain the popularity of selfie sticks.

TAKE COOLER SELFIES BY MESSING WITH YOUR FIELD OF VIEW

Perspective distortion can change the relative size of parts of your face, but there's another way it can affect your photos—one that can open up a whole new variety of selfie options.

When you zoom in, you change the apparent size of objects in the background. If you're standing in front of a large object that's far away, such as a mountain, the camera's zoom can dramatically affect how big the mountain looks.

If you set up your camera on a timer and walk far away from it, you can make even a fairly small mountain look huge.

 HERE'S ME VISITING THE MOUNTAINS!

 AREN'T THOSE JUST THE PILES OF DIRT OVER AT THE LANDFILL?

YEAH! I SET UP BASE CAMP NEAR THE OLD WASHING MACHINES.

MOON SELFIE

Smartphone cameras have limits to how far they can zoom, but if you get a camera with a powerful telephoto zoom lens, you can take some really interesting selfies. You can even recreate those Moon skyline photos – but using your own body instead of a building.

We can use geometry to work out how far away your camera needs to be in order to take a photo of yourself in front of the Moon.

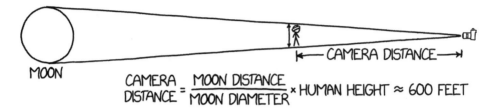

$$\text{CAMERA DISTANCE} = \frac{\text{MOON DISTANCE}}{\text{MOON DIAMETER}} \times \text{HUMAN HEIGHT} \approx 600 \text{ FEET}$$

This tells us that the camera needs to be about 600 feet away to take a Moon selfie.

OK, SMILE!

Since they don't make selfie sticks that are 600 feet long, you'll probably want to set up the camera on some kind of tripod and trigger it remotely.

Lining up a photo like this can be tricky; you need to find an area with a high place to stand and a long, unobstructed view path in the opposite direction from the Moon. The Moon moves quickly, so once everything is lined up, you'll only have a short window to take a photo—about 30 seconds. It only takes a little over two minutes for the Moon to move out of view completely.[1]

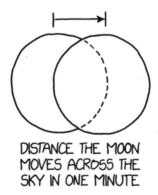

DISTANCE THE MOON MOVES ACROSS THE SKY IN ONE MINUTE

With the right filters, if you're extremely careful, you can even take a photo like this of the Sun. This may destroy your camera, so consult with your local astronomy club or photography store before trying this yourself. If you don't, there's a good chance you'll set

---

1   Tools like Google Earth and sky chart apps like Stellarium and Sky Safari can help you plan out the shot.

your camera on fire. And *never* look through an optical viewfinder when you're pointing a camera at the Sun. Your eye may not work exactly like a camera, but it's just as easy to burn a hole in.

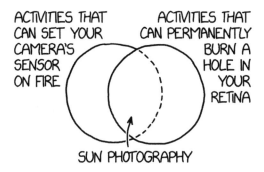

VENUS/JUPITER SELFIE

In principle, you can take a similar photo using even smaller, more distant objects. After the Sun and Moon, the celestial bodies that appear largest in the sky are Jupiter and Venus, which are both around an arcminute in size when close to the Earth and most visible. Using the same geometry from the Moon example, you can work out how far away you'll need to hold the camera in order to take a selfie with Venus or Jupiter: about 4 miles.

Holding a camera 4 miles away will present some obvious challenges.

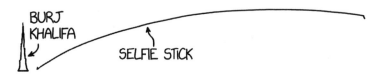

Atmospheric distortion is greatest when Venus is closest to the horizon, so you'll want it to be relatively high in the sky—which means you'll need to be high above the camera. But you want the camera to be fairly high up as well, to get it out of the thick atmosphere.

A good setup would consist of a camera on a mountaintop, with the subject standing on a much higher mountaintop. But finding two climbable mountains the right distance apart that align with Venus on a particular day will take a lot of survey work and planning. You could try to avoid the alignment problem by positioning yourself on a high-altitude aircraft or balloon, but maneuvering to get yourself in the right position will be extremely difficult and will probably require computer control.

Regardless of which method you choose, getting the alignment right will be an extremely difficult challenge, and whatever picture you take is going to be pretty blurry. Even under the best of conditions, it's difficult to take a sharp picture of Jupiter or Venus from the ground due to atmospheric distortion. It's possible no one has managed to take a selfie like this, so if you do, you'll definitely earn internet bragging rights.

A selfie with Jupiter or Venus would push the bounds of optics and geometry, and would be pretty hard to top . . . from Earth, that is. If you travel to space, where atmospheric distortion is less of a problem, you can open up new selfie possibilities.

There are several telephoto cameras in space with very high angular resolution, although you may have trouble convincing NASA to let you borrow them.[2]

But there's a way to take a space selfie with an even longer "zoom" than the fanciest space telescope. It's called *occultation*, and it's one of the coolest tricks in astronomy.

OCCULTATION SELFIES

When an asteroid passes in front of a star from the point of view of Earth, people with stopwatches scattered around the world can time when the star disappears and reappears, and use those measurements to build up a picture of the asteroid.

OBSERVERS ON EARTH

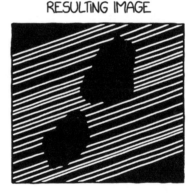

RESULTING IMAGE

This technique can be used to see detail too small or faint for the fanciest telescopes to make out. And it could, in theory, let you take an incredibly distant selfie while in space. All you need is a network of friends on the ground to watch a distant star blink out as you drift in front of it.

---

2   In theory, by the time this book is published, the James Webb Space Telescope should have finally launched. *[Editor's note: It was delayed again while this chapter was being edited.]*

Using a distant star, your friends could take a photo of you from a distance of up to several hundred miles. You won't be able to go any farther than that because your shadow will be lost to diffraction. If you use a distant X-ray source instead of a visible star, the shorter wavelength will reduce the effects of diffraction, and you could conceivably take a picture of yourself standing on the surface of the Moon while your friends observed from the ground.

Just remember: orbital alignments used for occultations are rare and usually don't repeat, so they take a huge amount of planning—which means you'll only get one shot.

WAIT, I WAS HOLDING MY ARM WEIRD.
CAN WE DELETE IT AND TRY AGAIN?

NO!!

CHAPTER 22

How to Catch a Drone

(with sports equipment)

A wedding-photography drone is buzzing around above you. You don't know what it's doing there and you want it to stop.

Let's assume you don't have any kind of sophisticated anti-drone equipment, like net launchers, shotguns, radio jammers, mist nets, counter-drone drones, or other such specialized gear.

If you *do* have a very well-trained bird of prey, you may think it's a good idea to send it after the drone. Every now and then, videos circulate around the internet showing trained birds of prey snatching drones out of the sky. This is a concept we find instinctively satisfying, but any plan that calls for countering rogue machines by training animals to hurl themselves at them is probably a bad one. We wouldn't enforce speed limits by training cheetahs to leap onto motorcycles. It would be cruel and dangerous to the cheetahs, and besides, there are a lot more motorcycles than there are cheetahs. Earth's MOTORCYCLE : CHEETAH ratio (MpC) has never been precisely calculated, but it's probably several hundred thousand.

Similarly, there are certainly more drones in the world than birds of prey—and new drones are being produced a lot faster than new birds of prey. Earth's DRONE : HAWK ratio is harder to estimate than its MpC, but it's almost certainly greater than 1.

MPC: 100,000+   DPH: ≧1

If falcons are a bad idea, what else might you use?

Drones are up in the air, so you want to send an object flying through the air. Humans send objects flying through the air all the time in the world of sports—see chapter 10: How to Throw Things for instructions.

Let's suppose you have a garage full of sports equipment—baseballs, tennis rackets, lawn darts,[1] you name it. Which sport's projectiles would work best for hitting a drone? And who would make the best anti-drone guard? A baseball pitcher? A basketball player? A tennis player? A golfer? Someone else?

I CAN CALL MY COUSIN; SHE'S A PRO GOLFER.

MY BROTHER DOES ARCHERY.

MY MOM ONCE THREW A HARPOON AT AN AUTOMATED MOVIE TICKET KIOSK.

There are a few factors to consider—accuracy, weight, range, and projectile size.

---

1 For those who weren't around in the 1980s, lawn darts were big heavy plastic darts with metal tips, similar to medieval weapons, which were sold to children as part of a game that involved throwing the darts high into the air. They were eventually banned in the United States for reasons that seem pretty obvious in retrospect.

A lot of drones are pretty fragile, so let's assume for the moment that as long as you can hit it, you'll cause it to crash. (This has certainly been my experience.)

For the purpose of approximate comparisons, we'll use a simple number to rate the accuracy of projectiles across different sports, representing the ratio of *range* to *error*. If you throw a ball at a target 10 feet away, and you miss by an average of 2 feet, then you have an *accuracy ratio* of 10 divided by 2, or 5.

The body of a medium-size drone—like the DJI Mavik Pro—has a "target area" about a foot across, meaning we can miss the center of the drone by 6 inches in either direction. If it's hovering 40 feet away, we'll need an accuracy ratio of 80 to be likely to hit it—or somewhat less if the projectile is larger, since that gives us more room for error.

Shots in which the projectile travels in a high arc, like in basketball or golf, gain additional accuracy, since the drone's wide and flat shape presents more of a target. And large projectiles, like footballs and basketballs, have more margin of error.

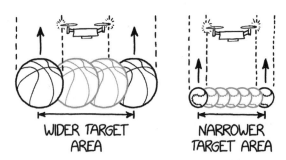

WIDER TARGET
AREA

NARROWER
TARGET AREA

Here are some estimated accuracy ratios for various athletes, based on competition play, exhibitions, or scientific studies in which athletes tried to hit targets.

Athlete	Estimated accuracy ratio	Attempts required to hit DJI Mavic Pro at 40 feet	Based on
Soccer kicker	21	13	Study of 20 experienced Australian players
Placekicker	23	15	NFL kickers in the late 2010s
Recreational hockey player	24	35	25 recreational and university hockey players
Basketball (Shaquille O'Neal)	36	4	NBA free throw percentage
Golf drive/chip	40	6[2]	PGA drive accuracy stats
Basketball (Steph Curry)	63	2	NBA free throw percentage
NHL All-Star	50	9	NHL accuracy shooting
NFL QB passing	70	4	Pro Bowl precision passing typical score[3]
High school pitcher	72	3	Study of 8 Japanese high school pitchers
Professional pitcher	100	2	
Darts champion	200-450	1[4]	PDC analysis of Michael van Gerwen
Olympic archer	2,800	1	2016 Korean men's archery team

[2] This is for a very accurate long drive. Accuracy for shorter-range chips may be higher.

[3] Quarterback Drew Brees, on the show *Sport Science*, threw a football at an archery target 20 yards away, hitting the bulls-eye ten times out of ten. This suggests that under those circumstances, his accuracy ratio is somewhere above 700—better than a darts champion.

[4] If they could keep their accuracy at that long range

Clearly, archers are the best choice, if you can find one. Their combination of extreme precision and long range would make them ideal defenders. Pitchers would also be a great choice—and a baseball would probably do a lot of damage. Basketball players make up for their lower accuracy with a large projectile and efficient arcing shot. Hockey players, golfers, and kickers are all probably less than ideal choices.

I was curious to test this in the real world, and one sport I couldn't find good data on was tennis. I found some studies of tennis pro accuracy, but they involved hitting targets marked on the court, rather than in the air.

So I reached out to Serena Williams.

To my pleasant surprise, she was happy to help out. Her husband, Alexis, offered a sacrificial drone, a DJI Mavic Pro 2 with a broken camera. They headed out to her practice court to see how effective the world's best tennis player would be at fending off a robot invasion.

The few studies I could find suggested tennis players would score relatively low compared to athletes who threw projectiles—more like kickers than pitchers. My tentative guess was that a champion player would have an accuracy ratio around 50 when serving, and take 5-7 tries to hit a drone from 40 feet. (Would a tennis ball even knock down a drone? Maybe it would just ricochet off and cause the drone to wobble! I had so many questions.)

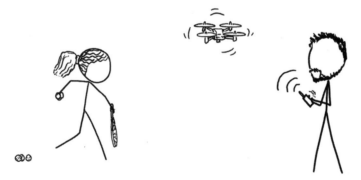

Alexis flew the drone over the net and hovered there, while Serena served from the baseline.

Her first serve went low. The second zipped past the drone to one side.

The third serve scored a direct hit on one of the propellers. The drone spun, momentarily seemed like it might stay in the air, then flipped over and smashed into the court. Serena started laughing as Alexis walked over to investigate the crash site, where the drone lay on the court near several propeller fragments.

I had expected a tennis pro would be able to hit the drone in five to seven tries; she got it in three.

Even though it's just a machine, a drone lying on the ground seems oddly tragic.

"I felt really bad hurting him," Serena said, after the pieces had been collected. "Poor little guy."

I couldn't help but wonder: Is it wrong to hit a drone with a tennis ball?

I decided to ask an expert. I contacted Dr. Kate Darling, robot ethicist at the MIT Media Lab, and asked her if it's wrong to hit tennis balls at a drone for fun.

She said, "The drone won't care, but other people might." She pointed out that while our robots obviously don't have feelings, we humans do. "We tend to treat robots like they're alive, even though we know they're just machines. So you might want to think twice about violence towards robots as their design gets more lifelike; it could start to make people uncomfortable."

That made sense, but on the other hand, should we really be making ourselves so vulnerable?

"If you're trying to punish the robot," she said, "you're barking up the wrong tree."

She has a point. It's not the robots we need to worry about, it's the people controlling them.

If you want to bring down a drone, perhaps you should consider a different target.

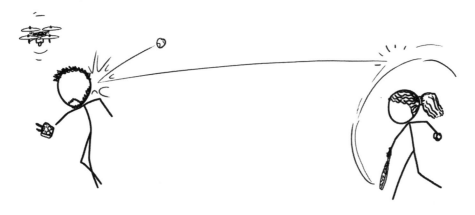

CHAPTER 23

How to Tell If You're a Nineties Kid

When were you born?

For most people, this is an easy question. Even those who don't know their exact birthday can usually tell when they were born to within a few years.

Yet the internet is full of quizzes that promise to help you determine which decade you were born in. These are usually based on what was happening in American pop culture at the time you first became aware of it.

THE APPLICATION FORM ASKED FOR MY BIRTHDAY, SO I'M TAKING A QUIZ TO FIND OUT IF I'M A NINETIES KID. THEN HOPEFULLY I CAN FIND ANOTHER QUIZ TO NARROW IT DOWN FURTHER.

Of course, these quizzes aren't really about helping you find out when you were born. They're about giving you a sense that there's a group out there that you belong to, reinforced by shared memories.

Movies and TV shows aimed at children are particularly well suited for this kind of quiz, not only because childhood memories are a source of nostalgia, but also because kids' shows often target a very limited age range, producing narrow "generational" distinctions. The mix of media you grew up with is often a unique fingerprint that shows your age down to within a few years. People born in the early to mid-1980s, for example, might remember the early "Disney Renaissance" movies as particularly formative – *The Little Mermaid* (1989), *Beauty and the Beast* (1991), and *Aladdin* (1992). On the other hand, those born in the *late* 1980s might have more-vivid, formative memories of *The Lion King* (1994) and *Toy Story* (1995). People born in the early 1980s were largely too old for the late-1990s Pokémon craze, while those born in the late 1980s were too young to listen to New Kids on the Block.

Clearly, there's a demand for these kinds of roundabout ways to identify your age. But why stop at movies and TV shows? The world is changing all the time in ways that leave a mark on us.

CHICKEN POX PARTIES

Chicken pox is an itchy rash, caused by the varicella-zoster virus, which lasts for a few weeks. After someone is infected once, they generally become immune to new infections for life (although the latent infection can flare up later in life, causing a painful rash called shingles).

For most of the 20th century, virtually everyone got chicken pox by the time they reached adulthood. Since the disease is more severe in adults than children, parents preferred to let their kids get exposed early – by throwing "chicken pox parties" – to gain immunity and avoid a risky later-in-life infection. Then everything changed[1] in 1995, when a chicken pox vaccine became available.

In the 10 years following its introduction, chicken pox vaccination rates climbed to near 100 percent, and chicken pox cases plummeted.

---

[1] If you expected the words "... when the Fire Nation attacked!" here, you are of a very specific age.

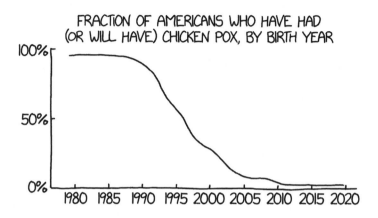

FRACTION OF AMERICANS WHO HAVE HAD
(OR WILL HAVE) CHICKEN POX, BY BIRTH YEAR

Within 20 years of the introduction of the vaccine, chicken pox went from universal to rare. For those born in the United States after the mid-1990s, it's seen as an old-timey disease, like polio. If you remember chicken pox and chicken pox parties, you were probably born in the 1990s or earlier.

SCARS

Chicken pox occasionally leaves lasting scars, but the chicken pox vaccine typically does not. Other vaccines left a physical mark on the generations that received them.

Smallpox, a disease caused by another virus, may well be responsible for the largest death toll of any human infectious disease. When Europeans arrived in the Americas, they brought smallpox with them—along with other diseases like hepatitis—into an area where the people had no natural immunity and were especially susceptible. The diseases swept across the continent, killing most of the people who lived there. The true death toll from smallpox is unknown, but it killed hundreds of millions of people in the 20th century alone.

Without its human hosts, the virus cannot survive. The first vaccine for smallpox was developed at the end of the 18th century, and by the end of the 19th century, the disease had become comparatively rare in most industrialized countries. In the 20th century, medical advances made the vaccine easier to produce and transport around the world, leading to a global campaign to eradicate smallpox completely. It succeeded: the last smallpox infection "in the wild" occurred in Somalia in 1977, and the last outbreak in history—and the final smallpox death—happened after a lab accident in 1978.

The smallpox vaccine is administered with a double-pointed needle, which breaks the skin in several places and deposits the vaccine:

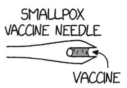

The vaccine contains a milder virus that causes the body to react as it would to a real smallpox infection, resulting in swelling, a blister, and a scab. After a few weeks, the wound heals, leaving a distinctive round scar.

The last case of smallpox in the United States was in 1949, and routine childhood smallpox vaccination in the United States and Canada ended in 1972.

If you're from the United States or Canada and have that vaccination mark on your upper arm or outer leg, it means you were born before about 1970.[2] The circular mark is a battle scar from humanity's war against one of our most terrible foes. And if you don't carry such a scar, that's a testament to our victory.

YOUR NAME

Baby names generally rise and fall in popularity over time.

Some names are relatively timeless; Elizabeth, Marshall, Susanna, Nina, and Nelson have been consistent in popularity in the United States over many generations. Biblical names, like John, James, and Joseph, are also relatively timeless. But changes in naming patterns can really sneak up on you. The biblical name Sarah was one of the most popular names in the United States in the 1980s, but by the mid-2010s there were more babies in the United States being named "Brooklyn" than "Sarah."

---

2   Vaccination continued in a few populations for a decade or so after routine childhood vaccination ended, mainly among people such as health care workers or soldiers who were considered to be at higher risk of infection.

Here's a list of some of the most common generationally specific names for every five years. These are names that had a relatively narrow peak in popularity, within just a decade or so. If you were born in the United States around this year, these are names that are more likely to seem common and generic to *you*, but are distinctive generational markers.

1880	*Will, Maude, Minnie, May, Cora, Ida, Lula, Hattie, Jennie, Ada*
1885	*Grover, Maude, Will, Minnie, Lizzie, Effie, May, Cora, Lula, Nettie*
1890	*Maude, May, Minnie, Effie, Mabel, Bessie, Nettie, Hattie, Lula, Cora*
1895	*Maude, Mabel, Minnie, Bessie, Mamie, Myrtle, Hattie, Pearl, Ethel, Bertha*
1900	*Mabel, Myrtle, Bessie, Mamie, Pearl, Blanche, Gertrude, Ethel, Minnie, Gladys*
1905	*Gladys, Viola, Mabel, Myrtle, Gertrude, Pearl, Bessie, Blanche, Mamie, Ethel*
1910	*Thelma, Gladys, Viola, Mildred, Beatrice, Lucille, Gertrude, Agnes, Hazel, Ethel*
1915	*Mildred, Lucille, Thelma, Helen, Bernice, Pauline, Eleanor, Beatrice, Ruth, Dorothy*
1920	*Marjorie, Dorothy, Mildred, Lucille, Warren, Thelma, Bernice, Virginia, Helen, June*
1925	*Doris, June, Betty, Marjorie, Dorothy, Lorraine, Lois, Norma, Virginia, Juanita*
1930	*Dolores, Betty, Joan, Billie, Doris, Norma, Lois, Billy, June, Marilyn*
1935	*Shirley, Marlene, Joan, Dolores, Marilyn, Bobby, Betty, Billy, Joyce, Beverly*
1940	*Carole, Judith, Judy, Carol, Joyce, Barbara, Joan, Carolyn, Shirley, Jerry*
1945	*Judy, Judith, Linda, Carol, Sharon, Sandra, Carolyn, Larry, Janice, Dennis*
1950	*Linda, Deborah, Gail, Judy, Gary, Larry, Diane, Dennis, Brenda, Janice*
1955	*Debra, Deborah, Cathy, Kathy, Pamela, Randy, Kim, Cynthia, Diane, Cheryl*
1960	*Debbie, Kim, Terri, Cindy, Kathy, Cathy, Laurie, Lori, Debra, Ricky*
1965	*Lisa, Tammy, Lori, Todd, Kim, Rhonda, Tracy, Tina, Dawn, Michele*
1970	*Tammy, Tonya, Tracy, Todd, Dawn, Tina, Stacey, Stacy, Michele, Lisa*
1975	*Chad, Jason, Tonya, Heather, Jennifer, Amy, Stacy, Shannon, Stacey, Tara*
1980	*Brandy, Crystal, April, Jason, Jeremy, Erin, Tiffany, Jamie, Melissa, Jennifer*
1985	*Krystal, Lindsay, Ashley, Lindsey, Dustin, Jessica, Amanda, Tiffany, Crystal, Amber*
1990	*Brittany, Chelsea, Kelsey, Cody, Ashley, Courtney, Kayla, Kyle, Megan, Jessica*
1995	*Taylor, Kelsey, Dakota, Austin, Haley, Cody, Tyler, Shelby, Brittany, Kayla*
2000	*Destiny, Madison, Haley, Sydney, Alexis, Kaitlyn, Hunter, Brianna, Hannah, Alyssa*
2005	*Aidan, Diego, Gavin, Hailey, Ethan, Madison, Ava, Isabella, Jayden, Aiden*
2010	*Jayden, Aiden, Nevaeh, Addison, Brayden, Landon, Peyton, Isabella, Ava, Liam*
2015	*Aria, Harper, Scarlett, Jaxon, Grayson, Lincoln, Hudson, Liam, Zoey, Layla*

If kids in your class were named Jeff, Lisa, Michael, Karen, and David, then you were probably born in the mid-1960s. If they were named Jayden, Isabella, Sophia, Ava, and Ethan, then you were probably born somewhere around 2010.

But names can reveal things about age in other ways.

The mid-1990s TV show *Friends* featured six roommates, played by actors, named Matthew, Jennifer, Courtney, Lisa, David, and another Matthew. Each of those names has its own popularity curve; if we combine them all, we can guess what years the group of actors was likely born:

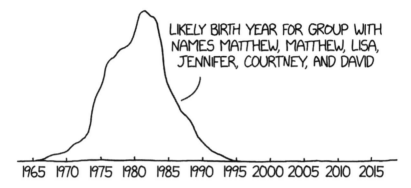

LIKELY BIRTH YEAR FOR GROUP WITH NAMES MATTHEW, MATTHEW, LISA, JENNIFER, COURTNEY, AND DAVID

The actors were actually born in the late 1960s, on the very early edge of the popularity of their names. In other words, the actors all have names that were a little before their time. Courtney Cox and Jennifer Aniston had names that didn't really become popular until a decade later. (Maybe people with trendy parents are more likely to wind up in acting.) But the names are generally consistent with their era, if a little ahead of the curve.

We get something very different if we look at the names of their *characters*—Phoebe, Joseph, Ross, Chandler, Rachel, and Monica:

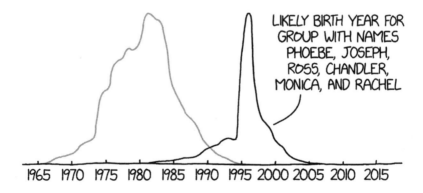

LIKELY BIRTH YEAR FOR GROUP WITH NAMES PHOEBE, JOSEPH, ROSS, CHANDLER, MONICA, AND RACHEL

The show debuted in 1994. There's a clear spike in popularity of the names in 1995 and 1996, which can probably be attributed to the show putting the names in the minds of new parents. But it's not just the show—that name combination was clearly on the rise in the years before *Friends* premiered. It's possible that parents looking for good names for their children are influenced by some of the same cultural trends as TV writers looking for good names for their characters.

RADIOACTIVE TEETH

Humans invented nuclear weapons in 1945. We detonated the first one to test whether they worked, then used another two in war. Once that war was over, we set off a few thousand more of them just to see what would happen.

We learned a lot about nuclear weapons from these tests. One thing we learned was, "Setting off nuclear weapons fills the atmosphere with radioactive dust." We also learned that nuclear weapons could be made much more powerful. In fact, there was effectively no limit on how powerful we could make them, which was a little alarming. The United States and the Soviet Union quickly developed arsenals large enough to, effectively, end the world. The knowledge that distant humans could trigger a fiery apocalypse at any moment at the push of a button left a strong impression on the children of the 1950s and 1960s.

But the impression it left was physical as well as psychological.

Most of the atmospheric nuclear explosions happened in the mid- to late 1950s, with a few more truly gigantic ones in 1961 and 1962. Amid increasing concerns about radioactive contamination, the United States and USSR agreed to stop all above ground testing and limit themselves to underground tests. They signed the Limited Nuclear Test Ban Treaty in 1963, which ended the era of large-scale atmospheric nuclear tests. Over the next few decades, there were only a few more atmospheric nuclear tests by France and China. The final nuclear explosion in the Earth's atmosphere was a Chinese test on October 16, 1980.[3]

The radioactive debris released by these explosions spread throughout the atmosphere. It consisted of a wide variety of radioactive elements. Some, like cesium-137, accumulated in human bodies and caused cancer. Others, like carbon-14, were harmless to human health, but caused annoyance to archaeologists by messing with carbon dating.

Carbon-14 is naturally produced by cosmic rays interacting with the atmosphere, and decays into nitrogen-14 with a half-life of about 5,700 years. At any given time, a tiny fraction of the carbon in the atmosphere is carbon-14; the rest is carbon-12 and carbon-13. Other than its limited lifespan, carbon-14 acts just like its stable cousins, and is incorporated into organic[4] material without causing problems. When an organism dies, its bio-

---

3 I don't know what year you're reading this sentence; I hope it's still true.
4 "Organic" means "carbon-based," after all!

logical processes stop exchanging carbon with the atmosphere, and the carbon-14 starts to decay. By measuring how much carbon-14 is left in an archaeological specimen, we can determine how long ago it stopped getting a fresh supply of carbon-14. In other words, we can figure out when it died.

This trick—carbon dating—is only possible if we know the original concentration of carbon-14 in the atmosphere when the organism was alive. Since carbon-14 is produced by cosmic rays, its concentration seems to have been relatively stable over time ... until we came along. Nuclear testing injected a huge amount of carbon-14 into the atmosphere:

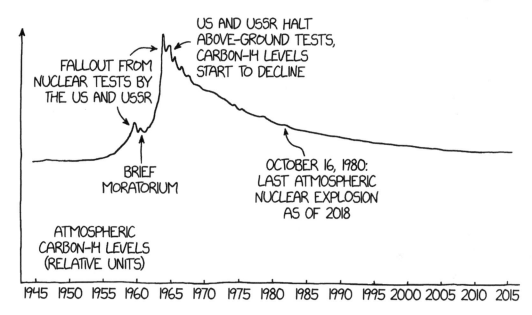

Any future archaeologists trying to carbon-date organic specimens will need to account for the massive 20th-century spike, or they'll calculate the wrong date for everything they dig up.

THESE ARE THE BONES OF HUMAN MUSICIANS WHOSE NAME TRANSLATES TO "THE NEW CHILDREN IN THE NEIGHBORHOOD." THEY PERFORMED IN THE 1990S, BUT CARBON DATING INDICATES THAT THE MEMBERS SURVIVED FOR NEARLY EIGHT CENTURIES.

Another contaminant released by nuclear testing was strontium-90. Since strontium is similar to calcium, our bodies incorporate it into teeth and bones. People who were children in the 1960s absorbed a lot of strontium. Researchers collected baby teeth throughout the 1950s and 1960s[5] and tested them for strontium-90, confirming the contamination and helping to make the case for a moratorium on atmospheric testing.

Atmospheric levels of strontium-90 declined after the early 1960s. Over time, the high strontium levels in the skeletons of the baby boomers fell, as the strontium was removed by the natural process of bone renewal. By the 1990s, baby boomers and children had similar levels of strontium in their bones.

Teeth, on the other hand, are more compact and stable than bones, and don't undergo natural renewal at the same rate that bones do. People whose permanent teeth were forming in the early 1960s likely carry ever-so-slightly elevated strontium-90 levels to this day.

In the same way that nuclear testing filled the atmosphere with radioactive fallout, the burning of leaded gasoline contaminated the air with lead. This led to a mid-20th-century lead poisoning epidemic, which peaked around 1972. The median childhood blood lead level in the late 1970s was 15 micrograms per deciliter, and it was probably even higher earlier in the decade. Children in many areas had lead levels above 20 µg/dL, levels that we now know to cause significant harm to developing brains. Studies suggest that lead in tooth enamel isn't exchanged with the environment, so baby boomers and Gen-Xers likely have elevated amounts of lead in their permanent teeth as well. These trace amounts of strontium and lead are now too small to have any real health effects, but we carry them with us as souvenirs.

Most of the contaminants from the mid-20th century are fading from the environment. Elements like iodine-131 emit a lot of radiation in the first few months but quickly decay.

---

5   At least, I *hope* those were researchers.

The longer-lived carbon-14 is being removed by the natural carbon cycle, and carbon-14 is almost back to its "natural" levels.[6] Strontium-90 has a half-life of around 30 years, as does cesium-137, another major source of contamination. As of the publication of this book, a quarter of the strontium-90 and cesium-137 from the 1960s nuclear tests remains.

But even as the radioactive elements settle out of the environment and slowly decay into more-inert forms, their imprint is left on us. No one really knows how many people have died from cancer caused by nuclear testing. The low estimates are in the thousands. The high estimates are in the hundreds of thousands. The quiet, hidden death toll from these weapons tests may well be greater than the death toll from the bombings of Hiroshima and Nagasaki. The legacy of the choices we made in that short period after World War II will be with us for a long time.

So if you want to know whether you're a 1990s kid or a 1950s kid, check your teeth.

GOT SOME TEETH.

ARE THOSE FROM BABY BOOMERS?

YEAH, BUT—

NO WAY, GET 'EM OUT OF HERE.

---

6 Burning of fossil fuels releases more carbon-12 and carbon-13 into the atmosphere, which actually reduces the carbon-14, but this effect is dwarfed by the huge increases caused by nuclear testing.

CHAPTER 24

How to Win an Election

THIS IS AN ELECTION, NOT A POPULARITY CONTEST.

AN ELECTION IS *LITERALLY* A POPULARITY CONTEST.

To win an election, you have to convince lots of people to choose your name on a ballot. There are two general approaches you could follow:

- Convincing lots of voters to support you
- Tricking them into picking your name on the ballot by mistake

The first approach typically requires some combination of charm, personal charisma, competence, a compelling message, and the presentation of a clear choice between competing visions for the future. Those are all a lot of work, so let's start by considering the second one.

TRICK VOTERS INTO
PICKING YOUR NAME BY MISTAKE

This electoral strategy has been popular over the years, despite the generally mixed results.

In 2016, a Canadian man from Thornhill, Ontario, spent $137 to legally change his name to "Above Znoneofthe," and filed to run in a provincial election. He intended to be listed on the ballot as "ZNONEOFTHE ABOVE," with the Z included to put him at the end of the alphabetical list of names, hoping that people would mistake his name for a "None of the above" option. Unfortunately for him, while the names on the ballot were alphabetized by last name, they were printed in <first name> <last name> order. He appeared in the ballot as "Above Znoneofthe." Mr. Znoneofthe did not prevail.

If you're running for a minor local office, it's possible that a large portion of voters won't know who you are, especially if you run in a year when there's a big election that draws lots of turnout.[1] In these situations, many voters might have nothing to judge you by except your name.

Occasionally, this has caused confusion — and created opportunities. In 2018, Kansas congressman Ron Estes ran for reelection, only to be challenged in the Republican primary by a political newcomer who was also named Ron Estes.

The second Ron Estes was listed on the ballot as Ron M. Estes. The incumbent Ron changed his campaign signs to list him as "Rep. Ron Estes" and ran ads informing voters that the *M* stood for "Misleading." The other Ron retaliated by telling voters it stood for "'Merica."

In the end, the name gimmick didn't work. When the primary finally arrived, Ron Two was soundly defeated by Ron Prime.

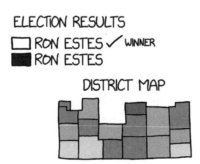

ELECTION RESULTS
☐ RON ESTES ✓ WINNER
■ RON ESTES

DISTRICT MAP

---

1   If people are showing up to vote for an exciting presidential candidate, they might not be as familiar with the rest of the candidates on the ballot as the more civically engaged people who vote in a midterm.

But name games *have* occasionally been successful – just ask Bob Casey.

From 1960 into the 21st century, Pennsylvania has voted for *five different people named Bob Casey* in statewide or federal elections, and it's not entirely clear that voters always chose the Bob Casey they intended.

Here's a quick rundown of the Bob Caseys[2] of Pennsylvania:

- **Bob Casey #1:** a lawyer from Scranton
- **Bob Casey #2:** Cambria County Recorder of Deeds
- **Bob Casey #3:** a PR consultant
- **Bob Casey #4:** a schoolteacher and ice cream seller
- **Bob Casey #5:** son of Bob Casey #1

BOB CASEY #1 (LAWYER)   BOB CASEY #2 (COUNTY OFFICIAL)   BOB CASEY #3 (P.R. CONSULTANT)   BOB CASEY #4 (ICE CREAM SELLER)   BOB CASEY #5 (SON OF BOB CASEY #1)

Starting in the 1960s, Bob Casey #1 was elected to a number of state offices, quickly becoming a rising star in state politics. In 1976, the state held an election for treasurer. Bob Casey #1, then the state auditor, was eying a campaign for governor in 1978, so he decided not to run for treasurer ... but Bob Casey #2 – a Cambria County official – did.

The same year, Bob Casey #3 ran for Congress in Pennsylvania's 18th district. His opponent complained that he was just trying to take advantage of Bob Casey #1's popularity. Bob Casey #3 retorted that Bob Casey #2 was cashing in on the collective popularity of himself and Bob Casey #1, the *real* Caseys. Bob Casey #3 ended up winning the Republican nomination, but lost the general election to the Democrat.

As for Bob Casey #2, despite barely campaigning, he also won his primary, defeating Catherine Knoll – the party-endorsed candidate – and several others. Knoll's campaign spent $103,448; Casey spent $865.

---

2   Or is it "Bobs Casey"?

Casey went on to win the general election and serve a four-year term as treasurer. The Republicans began a campaign to inform the public that "Bob Casey" wasn't who they thought he was, and their nominee—Budd Dwyer—defeated Casey #2 in 1980.[3]

In 1978, during Bob Casey #2's term as treasurer, Bob Casey #1 launched a bid for governor. Unfortunately, this was the same year Bob Casey #4—a schoolteacher and ice cream seller from Pittsburgh—appeared on the scene. Bob Casey #1 ran for governor, but Bob Casey #4 ran for lieutenant governor in the same primary. Voters, possibly thinking that Bob Casey #1 was making himself available for both positions,[4] nominated Bob Casey #4 lieutenant governor, but chose Pete Flaherty over Bob Casey #1 for governor. In the end, the Flaherty-Casey #4 ticket lost the general election.

In 1986, Bob Casey #1 ran for governor again, branding himself as "The Real Bob Casey," and finally won.[5] He served for eight years as governor before leaving office in 1994. Two years later, Bob Casey #5—his son, Bob Casey Jr.—ran for auditor and won. He went on to become state treasurer and eventually senator, a position he was reelected to in 2018.

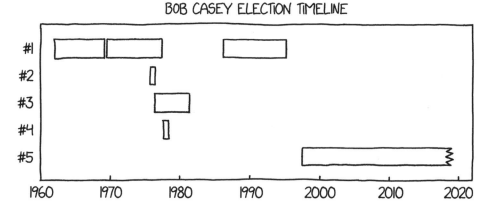

So if you're going to run for office, try changing your name to Bob Casey. You never know!

---

3   Catherine Knoll did end up being elected treasurer later, in 1988, and went on to also serve as lieutenant governor.

4   Or perhaps they thought that the state treasurer, Bob Casey #2, was running for governor midterm.

5   He won with the help of campaign strategist James Carville, who would later be part of Bill Clinton's successful presidential campaign.

CONVINCING LOTS OF VOTERS
TO SUPPORT YOU

Winning elections is hard. The truth is, people are complicated, there are a lot of them, and no one is ever 100 percent sure why they do what they do or what they're going to do next.

But if your goal is simply to win an election, then as a general rule you should be *for* things that voters like and *against* things they dislike. To do that, you'll need to figure out what the voters like and dislike.

One of the most popular tools for figuring out what the public thinks is opinion polling—talking to a bunch of people, asking them what they think, and tallying up the results.

The website FiveThirtyEight has conducted an exercise in which they had professional speechwriters write a speech that simply pandered as much as possible—only making statements that most voters support, to pander either to one party or to the electorate in general.

But what do we agree on the *most*? If your goal is simply to be in favor of popular things and against unpopular things, what should you campaign on? What are the *least* controversial issues in the country?

To help figure this out, I reached out to Kathleen Weldon, director of data operations and communications at the Roper Center for Public Opinion Research at Cornell University, to commission a poll of their polls. The Roper Center maintains a tremendous database of opinion polling data—over 700,000 polling questions spanning almost a century of opinion polling, collected from virtually every organization that has ever conducted a public poll in the United States.

I told them I was looking for the most one-sided questions in their polling database—the questions where virtually everyone gave the same answer. In a sense, these would be the least divisive issues in the country.

The Roper research staff sifted through their database of 700,000 questions and assembled a list of those questions for which at least 95 percent of respondents gave the same answer.

It's pretty rare for that many respondents to agree on *anything* in a poll. A small percentage of respondents will often choose ridiculous answers because they're not taking the poll seriously, or because they misunderstand the question. But one-sided questions are also rare because no one bothers to conduct polls on uncontroversial topics unless they're trying to prove a point. Since everything in the Roper database is something that some person or organization bothered to commission a poll to ask, it means it's at least *potentially* controversial, if not actually so.

Here is a selection of the most one-sided issues in the history of polling. If you want to run for office, these are views you can safely espouse, secure in the knowledge that at least one scientific survey puts the people squarely behind you:

Popular Opinions
According to actual data (Full question text in endnotes)

95% disapprove of people using cell phones in movie theaters.
(Pew Research Center's American Trends Panel poll, 2014)

97% believe there should be laws against texting while driving.
(*The New York Times* / CBS News Poll, 2009)

96% have a positive impression of small business. (Gallup Poll, 2016)

95% believe employers should not be able to access the DNA of their employees without permission. (*Time* / CNN / Yankelovich Partners Poll, 1998)

95% support laws against money laundering involving terrorism.
(ABC News / *Washington Post* Poll, 2001)

95% think doctors should be licensed. (Private Initiatives & Public Values, 1981)

95% would support going to war if the United States were invaded.
(Harris Survey, 1971)

96% oppose legalizing crystal meth. (CNN / ORC International Poll, 2014)

95% are satisfied with their friends. (Associated Press / Media General Poll, 1984)

95% say that "if a pill were available that made you twice as good looking as you are now, but only half as smart," they would not take it.
(Men's Health Work Survey, 2000)

98% believe lifeguards should watch swimmers rather than reading or talking on the phone. (American Red Cross Water Safety Poll, 2013)

99% think it's wrong for employees to steal expensive equipment from their workplace. (*Wall Street Journal* / NBC News Poll, 1995)

95% think it's wrong to pay someone to do a term paper for you.
(*Wall Street Journal* / NBC News Poll, 1995)

98% would like to see a decline in hunger in the world. (Harris Survey, 1983)

97% would like to see a decline in terrorism and violence. (Harris Survey, 1983)

98% would like to see a decline in high unemployment. (Harris Survey, 1982)

97% would like to see an end to all wars. (Harris Survey, 1981)

95% would like to see a decline in prejudice. (Harris Survey, 1977)

95% don't believe Magic Eight Balls can predict the future. (Shell Poll, 1998)

96% think the Olympics are a great sports competition.
(*Atlanta Journal Constitution* Poll, 1996)

You can use this list to assemble a campaign platform. For example, you could stand firmly against hunger, war, and terrorism; for friendship and small business; and against texting while driving. You could support laws that ensure doctors are properly licensed, and oppose allowing other countries to invade.

On the other hand, if you wanted to *lose* an election as spectacularly as possible, this list could be even more helpful as a blueprint. By taking the opposing position on each issue, you could potentially run the most *un*popular political campaign in political history. You'd probably lose, but in a world that has nominated at least five different Bob Caseys, who knows!

A VOTE FOR ME IS A VOTE FOR HIGH UNEMPLOYMENT, WAR, WORKPLACE THEFT, AND TEXTING WHILE DRIVING. I BELIEVE EVERY CITIZEN'S VOICE SHOULD BE HEARD IN EVERY MOVIE THEATER IN THE COUNTRY. IF ELECTED, I VOW TO PUT AN END TO THE OLYMPICS ONCE AND FOR ALL.

MY ADMINISTRATION WILL RAISE TAXES ON SMALL BUSINESS AND USE THE MONEY TO INSTALL A VIDEO GAME CONSOLE IN EVERY LIFEGUARD CHAIR IN THE COUNTRY. WE WILL MANUFACTURE AND SELL CRYSTAL METH, AND USE THE PROCEEDS TO GIVE TAX CREDITS TO ANYONE WHO PRACTICES MEDICINE WITHOUT A LICENSE. WE WILL ENGAGE IN MONEY LAUNDERING, BUT ONLY TO SUPPORT TERRORISM. EVERY DECISION IN MY ADMINISTRATION WILL BE MADE BY MAGIC EIGHT BALL. IF OUR COUNTRY IS INVADED, I WILL SURRENDER IMMEDIATELY.

VOTE FOR ME IF YOU LOVE HUNGER. VOTE FOR ME IF YOU HATE YOUR FRIENDS. AND IF YOU VOTE FOR ME, I PROMISE YOU THIS: EVERY ONE OF YOU WILL BE TWICE AS ATTRACTIVE AND HALF AS SMART.

CHAPTER 25

How to Decorate a Tree

About three-quarters of American households decorate trees for Christmas.

As of 2014, two-thirds of those households use artificial trees, while one-third use real live trees. The vast majority of people who use real trees get them from Christmas tree farms, but the traditional method – according to mid-20th-century Christmas films, anyway – is to simply walk out into the woods and find a suitable tree to cut down.

Depending where you are, you may not find a forest nearby. Forests are distributed a little irregularly around the world. Most of the world's forests are concentrated along the equator and in the polar latitudes. The forests of the equator and the poles are separated by bands of desert located around 30° north and south of the equator.[1] If you're near 30°N or 30°S, and you don't see any forests, try walking a few thousand miles toward the pole or the equator.

Once you locate a forest – and, ideally, obtain permission from the landowner – your next challenge is selecting a Christmas tree.

---

1   There are some exceptions to this. The US coast of the Gulf of Mexico is thickly forested, despite lying in the desert latitudes, thanks to the warm moist air from the Gulf. This is also why the area has so many tornadoes.

But be careful what tree you cut down.

In 1964, University of North Carolina graduate student Donald Currey was studying the history of glaciers in Nevada. A decade earlier, another scientist, Edmund Schulman, had discovered some very old trees nearby. Some of the bristlecone pines Schulman studied turned out to be between 3 and 5,000 years old—older than any other known trees.

Schulman's ancient trees were in California's White Mountains. Currey, over the border in Nevada, found bristlecone pines as well, and suspected they were of a similar age. He started sampling the trees, reasoning that their age could reveal something about the history of the ice age he was studying. If the area had cooled and the glaciers had expanded, the trees would have retreated down the mountain—so the pines at the uphill edge of the grove should be relatively young. He set about taking samples from some of the pines to determine their age.

Accounts differ about exactly what happened next. Literature professor and mountaineer Michael P. Cohen, in a 1998 book about the Great Basin, cataloged five different versions of the incident from people involved, each one spinning things a little differently.

All accounts agree on the central facts: Currey located a tree that seemed especially old (unbeknownst to him, local naturalists had dubbed it "Prometheus") and obtained permission from the Forest Service to cut it down to determine its precise age. After counting the rings from the trunk sections, Currey determined that the bristlecone pine was at least 4,844 years old, making it the world's oldest known tree.

When Currey's findings were published, public outcry ensued, and everyone involved in the project spent the next several decades trying to explain why they had killed the oldest tree on Earth.

LISTEN, THE TREE AND I WERE LOCKED IN MORTAL COMBAT. IT WAS IT OR ME!

The lesson of history is clear: before cutting down a tree, you may want to make sure it's not the oldest one in the world or people will get really mad.

Since the fall of Prometheus, the oldest dated tree has been another bristlecone pine, dubbed "Methuselah." Methuselah is at least 4,851 years old as of 2019, which means it has recently exceeded Prometheus's record.

These ages are determined from examining core samples, and so they only give a lower limit on the true age, since some of the youngest parts of the tree may not be represented in the core. University of Arizona researchers obtained pieces of Prometheus's trunk and determined that the tree was almost exactly 5,000 years old when it was cut down. That means it was . . . born? . . . hatched? . . . sprouted? . . . germinated? . . . around 3037 BCE, followed by Methusaleh some decades later. At the time the bristlecones first sprouted, humans on the other side of the world were developing the first known writing systems in Sumer.[2]

The forestry community is clearly hoping to avoid another Prometheus incident. Methuselah isn't exactly under round-the-clock armed guard, but its exact identity and location remains a secret, to protect it from damage from either souvenir hunters or — perhaps — copycat killers.

---

2   Another tree, dated by the late dendrochronologist Tom Harlan, may be slightly older than Methuselah or Prometheus. However, the tree's record is disputed — the organization Rocky Mountain Tree-Ring Research has been unable to locate the core to verify the age.

ONE OF THESE TREES IS IN THE WITNESS PROTECTION PROGRAM, BUT THERE'S NO WAY TO KNOW WHICH ONE.

These bristlecone pines are certainly unique, but the truth is, they would make terrible Christmas trees. You might think the trees that reach the greatest age are the ones that grow in the healthiest and most supportive environments. Surprisingly, the opposite is true. The oldest trees tend to be the ones that grow in the worst conditions, not the best ones. When a bristlecone pine is in a particularly harsh environment, one which blasts the trees with heat, cold, wind, and salt, it slows their growth and development, stretching out their lifespans. They weren't very impressive to look at—the very oldest among them looked like dead trees, with just a thin strip of bark running up one side, supporting a few branches that clung to life. These ancient trees aren't immortal—they've just figured out how to die slowly.

If the world's oldest tree would make a bad Christmas tree, what about the world's tallest?

US towns occasionally claim to host the world's tallest Christmas tree. According to *Guinness World Records*, that title belongs to a 67-meter Douglas fir, erected in 1950 at a shopping mall in Seattle. Of course, like all such seemingly trivial records, if you dig a little deeper you find a bitter controversy. In 2013, the *Los Angeles Times* published a story on large Christmas trees, in which tree farm owner John Egan accused the Seattle record of being bogus. Egan claims that the Seattle record holder was not a real tree; rather, it was constructed from several trees attached end-to-end. Egan claims that the real record holder is a 41-meter tree his own company erected in 2007.

Regardless of who the real record holder is, Egan points out that the record would be fairly easy to break. Someone just needs to go cut down a larger tree—and there are plenty that are bigger than either contender.

ONLY ONE THING CAN SETTLE
THE BITTER RIVALRY BETWEEN
THE NORTHGATE MALL AND
EGAN ACRES TREE FARM:

THEY MUST UNITE AGAINST
A COMMON ENEMY.

The world's tallest known tree is a coast redwood dubbed "Hyperion." Discovered in 2006, it stands just shy of 116 meters tall.[3] Bristlecone pines aren't the only record-setting tree in a witness protection program: Hyperion's exact location is kept quiet to protect it from harm, inasmuch as it's possible to hide something that tall.

ONE OF THESE IS THE TALLEST
TREE ON EARTH, BUT ITS IDENTITY
IS A CLOSELY GUARDED SECRET.

---

3   How do they measure a tree like that? You might think it involves GPS or lasers or something, but no: researchers just climb up and then dangle a tape measure down to the ground.

But there are a number of trees of similar height. Before Hyperion's measurement was revealed in 2006, the record holder was the 113-meter "Stratosphere Giant," another northern California coast redwood.

There are several other trees near Hyperion around that height, over 110 meters, and any one of them would work about as well as a Christmas tree. After all, who's going to get mad at you for cutting down the world's *second*-biggest tree?

DISPLAYING THE TREE

Where should you put up your tree? It's probably not going to fit in your house. In fact, there are very few buildings where it will fit.

The US Capitol rotunda (55 meters) and the highest stadium domes (around 80 meters) are too short to fit a coast redwood Christmas tree. Even the long halls of the biggest cathedrals, with nave heights of 40 or 50 meters, aren't tall enough. A Hyperion-size tree could just barely fit under the dome at St. Peter's Basilica at the Vatican, but only if you let the top poke into the lantern at the top of the dome.

ST. PETER'S BASILICA

In Halbe, Germany, southeast of Berlin, there's a former airship hangar that has been converted into a tropical theme park. It features hundreds of feet of sandy beaches, a rain forest section, and a water park. Unfortunately, the ceiling of the park is just a few meters too short to fit the tallest redwoods. You could still set up a tree there, but you'd need to dig out the floor first.

TROPICAL ISLANDS RESORT (AERIUM AIRSHIP HANGAR)

There are a few buildings with rooms large enough to hold a coast redwood Christmas tree. The Basilica of Our Lady of Peace, in Yamoussoukro, Côte d'Ivoire, is likely one of them. So are the atriums of several skyscrapers, including Dubai's Burj Al Arab (180 meters) and Beijing's Leeza SOHO (190 meters).

Even if the owners wanted to display your Christmas tree, getting the tree inside would be difficult, since the atriums of these buildings lack the necessary giant doors.

Perhaps the ideal building for displaying a giant Christmas tree is one found in the southeastern United States, on the east coast of Florida.

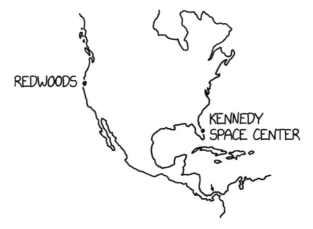

NASA constructed the Vehicle Assembly Building at Cape Canaveral as a place to prepare the Apollo rockets and Space Shuttles for launch. It's one of the largest buildings in the world by volume, with a ceiling easily high enough to fit your Christmas tree. And, crucially, there's a way to get the tree inside—the building has the tallest doors in the world.

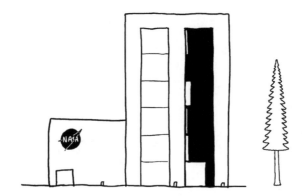

The easiest way to get it there will probably be by ship. Luckily, the Panama Canal is large enough to fit a boat carrying an intact 110-meter redwood lying on its side.

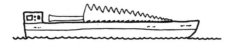

The VAB is a perfect fit for our tree for a simple reason: it was designed to hold the huge Saturn V rocket that took the Apollo astronauts to the Moon, and that rocket was almost exactly the same size as the world's tallest tree.

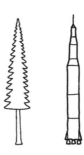

When fully loaded with fuel, the Saturn V rocket was substantially heavier than a Hyperion-size tree. Since the rocket's engines are capable of lifting it, that means that if you attached them to the tree, they could lift it, too.

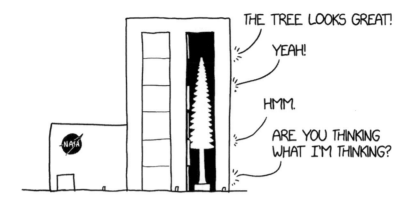

A pair of Space Shuttle booster rockets on either side would produce more than enough thrust to lift your tree.

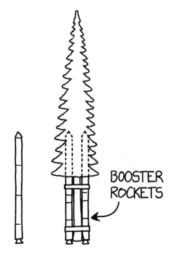

The tree itself would need some extra support. First, the tree will suffer from the extreme vertical acceleration. Redwoods, the tallest trees in the world, have to work to hold themselves up against gravity under the best of circumstances. A rocket launch might subject the tree to several additional g's of acceleration, doubling or tripling the apparent force of gravity and causing the tree to buckle.

You can make things easier for the tree by pulling it instead of pushing it. Wood, like many materials, is stronger in tension than compression. If you attach the booster rockets partway up the trunk, the wood on the bottom half will be under tension, since it's dangling behind the rocket, while only the top half will be under compression. By adding supports along the tree, you can help keep it steady and prevent it from collapsing.

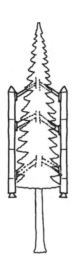

The rockets wouldn't be able to get the tree going fast enough to stay in orbit, but you could launch it on a suborbital trajectory that carried it over all those other towns that claim to have the largest Christmas tree.

And *your* tree would be decorated with actual stars.

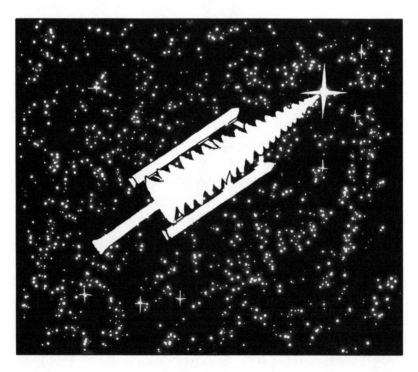

How to Build a Highway

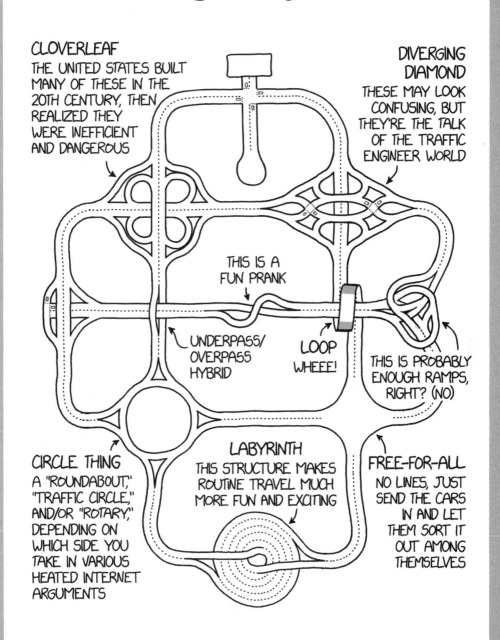

CLOVERLEAF
THE UNITED STATES BUILT MANY OF THESE IN THE 20TH CENTURY, THEN REALIZED THEY WERE INEFFICIENT AND DANGEROUS

DIVERGING DIAMOND
THESE MAY LOOK CONFUSING, BUT THEY'RE THE TALK OF THE TRAFFIC ENGINEER WORLD

THIS IS A FUN PRANK

UNDERPASS/ OVERPASS HYBRID

LOOP
WHEEE!

THIS IS PROBABLY ENOUGH RAMPS, RIGHT? (NO)

CIRCLE THING
A "ROUNDABOUT," "TRAFFIC CIRCLE," AND/OR "ROTARY," DEPENDING ON WHICH SIDE YOU TAKE IN VARIOUS HEATED INTERNET ARGUMENTS

LABYRINTH
THIS STRUCTURE MAKES ROUTINE TRAVEL MUCH MORE FUN AND EXCITING

FREE-FOR-ALL
NO LINES, JUST SEND THE CARS IN AND LET THEM SORT IT OUT AMONG THEMSELVES

CHAPTER 26

How to Get Somewhere Fast

Getting around the world can be incredibly complicated.

ACCORDING TO THIS NAVIGATION
APP, I NEED TO MOVE MY BODY
IN A COORDINATED AND VERY
SPECIFIC SERIES OF MOVEMENTS
TO PROPEL MYSELF FORWARD.

Depending on where you are and where you're going, you might be able to travel to your destination quickly in a relatively straight line, or you might need to go slowly and take an extremely roundabout route. Traveling can require solving a staggering variety of problems, from the basics of walking through doors to complicated tasks like navigating airport security, steering a car through rush-hour traffic, or plotting rocket engine maneuvers for orbital transfers.

But one way or another, traveling to a destination eventually involves accelerating yourself toward it. This acceleration puts a fundamental limit on how fast you can get where you're going.

Suppose you're trying to travel between Point A – say, your front yard – and Point B – say, a doctor's appointment – under totally ideal circumstances. There are no obstacles, no doors, no stop signs, and you have a magic scooter with unlimited fuel. How fast can you travel from Point A to Point B?

POINT A

POINT B

Everything on Earth is being accelerated downward by the pull of gravity at 9.8 m/s², or 1 G. When you accelerate forward in a vehicle, gravity is still pulling you downward, so the total acceleration you feel is the combination of the two – the horizontal push from the vehicle and the downward pull of gravity.

For small accelerations, the total acceleration you feel is about 1 G. If you accelerate at 0.1 G, the total acceleration you feel is just 1.005 G, but if you accelerate horizontally at 1 G, you feel a total acceleration of 1.41 G – as if every part of your body suddenly weighed 41 percent more.

Human transportation methods, from legs to elevators to cars to airplanes, usually involve horizontal accelerations of less than 1 G, for a few reasons. One big reason is that humans evolved to experience 1 G of acceleration, and we find it uncomfortable to spend much time accelerating faster than that. Another reason is that vehicles often accelerate by pushing against the ground, and when the horizontal push is stronger than the down-ward pull of gravity, the vehicle can find itself spinning its wheels.[1]

Let's assume your magic scooter is limited to accelerating at no more than 1 G. Real vehicles can and do accelerate faster than this on occasion, but generally they're specialty vehicles like rockets or roller coasters, and they only maintain that acceleration for a short period of time. If we're thinking about systems that the general public might use to get around, a 1 G scooter serves as a good model for the limits of what's possible while maintaining some semblance of human comfort and safety. Fighter pilots may be able to survive the sudden acceleration of an ejection seat with only minor injuries, but you wouldn't want to make one part of your commute.

---

1   Very fast sports cars accelerate at about 1 G, and need specialized high-grip tires to do so.

You hop on your scooter and check the clock. Your doctor's office is 500 meters away, and your appointment starts in 10 seconds. Can you make it? You turn up the throttle and accelerate toward the office.

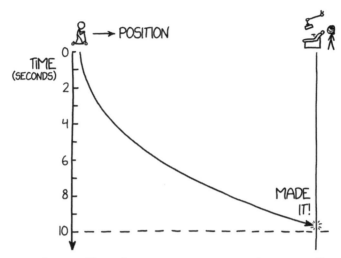

The good news is that you'll reach your appointment with many milliseconds to spare. The bad news is that you'll be traveling at over 200 mph when you arrive.

Unless your doctor doesn't mind very brief visits, you'll need to slow back down as you approach the office, which cuts into your total trip time. There are limits to how quickly you can decelerate; stopping is generally easier than starting—on virtually all land vehicles, from scooters to cars to taxiing airplanes, brakes are more powerful than propulsion—but stopping too suddenly causes just as many problems for your passengers as accelerating too fast.

If you spend the first half of your trip accelerating at 1 G and the second half decelerating at 1 G, it will take you almost 15 seconds to reach your appointment; if you leave 10 seconds before it starts, you won't get there in time.

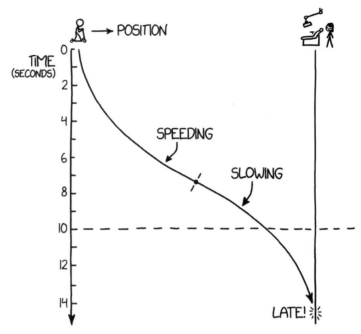

The limits faced by our magic scooter apply to any transportation method, from moving walkways to bullet trains to futuristic vacuum tube people movers, because they're a function of human biology. No transportation system will *ever* be able to carry people from a stationary position to a destination 500 meters away in less than 10 seconds without accelerating them horizontally harder than 1 G.

FUNDAMENTAL TRANSPORTATION RADIUS

1 SECOND 5 METERS 5 METERS

5 SECONDS 120 METERS 120 METERS

10 SECONDS 500 METERS 500 METERS

What if your appointment is farther away? How quickly can your scooter get you there?

At 1 G of continuous acceleration, speed adds up quickly. If you took a minute-long trip at 1 G of acceleration – speeding up for 30 seconds, slowing down for 30 seconds – you could travel more than 5 miles. Your peak speed, halfway through your trip, would be close to the speed of sound.

Real trains don't travel at near-supersonic speeds, but that's not due to some built-in limitation of physics. A platform on a rail can easily be accelerated up to extremely high speeds through electromagnetic propulsion or rockets. Rocket sleds running on rails at Holloman Air Force Base in New Mexico, for example, have traveled as fast as Mach 8, eight times the speed of sound—faster than any jet aircraft. To reach those speeds, the sleds accelerate much faster than 1 G—and even so, they require a test track nearly 10 miles long.

At speeds near the speed of sound, air resistance becomes an unavoidable problem—it's hard for a vehicle to be efficient when it's wasting so much energy pushing its way through the air. That's why the fastest vehicles tend to operate high in the atmosphere, where the air is thin, or in vacuum tubes. Your magic scooter, with its unlimited acceleration, doesn't face any of these problems, but hopefully it also has a good protective windshield. (You may also want to apologize to any bystanders for the sonic booms.)

In 5 minutes, a 1 G scooter could carry you 137 miles, reaching speeds of over Mach 4. In 10 minutes, you could travel 500 miles, reaching Mach 8. And in 48 minutes, you could travel halfway around the world.[2] This is the fundamental limit for world travel—if you want to build a system that shuttles people to anywhere in the world in under 48 minutes, it will have to involve accelerations of more than 1 G (or drilling a hole through the Earth).

---

2   Your actual travel time will be a little more complicated to work out, since at those speeds the curvature of the Earth becomes significant. Your speed in the middle of your journey would be fast enough that you'd lose contact with the ground, and if you tried to hold on to a rail (or drove on the ceiling), the centripetal acceleration would exceed your limits. But the same curvature means that you'd be able to accelerate a little harder near the beginning and end of your journey, since the centrifugal force helps cancel out the effect of gravity, giving you more leeway to accelerate forward while staying under the 1.41 G limit.

SPACE TRAVEL AT 1 G

These fundamental acceleration limits apply to spacecraft just as they do to Earth vehicles. If you outfit your magic scooter to leave the atmosphere and travel through space, accelerating and decelerating at 1 G, it will take you nearly 4 hours to make the trip to the Moon.

A 4-hour travel limit to the Moon tells us something interesting about the future. Even in a world with space elevators and cheap space travel, large numbers of humans living on Earth will probably not commute every day to the Moon—or vice versa—for simple reasons of acceleration. Four hours each way is a pretty long commute.

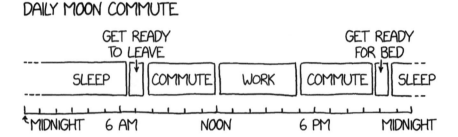

MORE DISTANT DESTINATIONS

It would take your 1 G scooter several days to reach any of the inner planets, a week to reach Jupiter, and 9 days to reach Saturn.

The outer planets Uranus and Neptune are about two weeks away, and reaching the more distant Kuiper Belt objects could take months.

Then things get weird.

AROUND THE UNIVERSE IN 80 YEARS

We don't have any spacecraft technology right now that can accelerate a vehicle at 1 G for long periods of time. There's nothing in physics that says it's impossible, but no one has figured out a way to do it; none of the energy sources we know of are small enough to be carried on a rocket yet powerful enough to accelerate it for that long. But if we ever *do* find a way to do it, it will open up the whole universe—thanks to a surprising boost from relativity. It turns out that if you accelerate at 1 G for several years, you can reach almost any destination in the universe.

If you accelerate at 1 G, your speed increases by 9.81 m/s every second. After 1 year, simple multiplication suggests you should be traveling at about 309 million m/s... which is 103 percent of the speed of light. Relativity tells us you can't really travel faster than light, so we know that's wrong—you can get closer and closer to the speed of light, but you can never quite reach it. Yet there aren't any cosmic police who show up and force you to stop accelerating, so what actually *happens* to you?

Strangely, from your point of view, nothing happens as your scooter approaches the speed of light. You just continue accelerating. But if you look out at the universe around you, you'll notice things getting a little strange.

As you go faster, the passage of time on board your scooter slows down. From the perspective of an outside observer, your scooter flies past carrying slowly ticking clocks and a slowly thinking brain. From *your* point of view, it seems to take you less time than it should to reach successive landmarks along your trip—as if the universe has contracted in the direction you're traveling.

After a year has passed for you on your scooter, you'll be traveling at about ¾ of the speed of light. But thanks to relativity, a year and two months will have passed in the outside world, and your ship will have flown farther than you'd expect.

The disparity between time on your scooter and time in the outside world keeps growing. After 1½ years have passed on board, you'll have traveled almost 1½ light-years, the same distance light would travel in that time. After 2 years have passed for you, you'll have gone *more* than 2 light-years, as if you've traveled faster than light!

After a few years have passed on board, the effects of relativity really start to add up. When 3 years have passed for you, a little over 10 years will have passed outside the ship, and you'll have traveled nearly 10 light-years—far enough to reach many nearby stars. If there were mile markers in space showing the distance that you've traveled, you'd be hitting them more and more quickly, as if they were closer and closer together—or as if you were traveling much faster than light. Yet to outside observers, you'd fly past at slightly less than the speed of light, everything on board apparently frozen in time.

After 4 years of scooting, you'll have gone 30 light-years, and you'll be traveling at 99.95 percent of the speed of light. After 5 years, you'll be 80 light-years from where you started, and after 10 years, you'll have traveled *15,000* light-years, bringing you halfway to the center of the Milky Way. If you continue accelerating, it would take you less than 20 years of your time to reach a neighboring galaxy.

If you keep pressing the accelerator for a little over two decades, you'll find your vehicle traveling billions of light-years per subjective "year," carrying you across a substantial fraction of the observable universe.

In that time, billions of years will have passed back home—so you don't need to worry about returning. The Earth will already have been consumed by the Sun anyway.

But you'll never reach the farthest galaxies. The universe is expanding, and thanks to dark energy, the expansion appears to be accelerating.

Traveling at nearly the speed of light may keep *you* from getting older, but the rest of the universe continues aging around you. If you travel a billion light-years at roughly the speed of light, the universe will be a billion years older when you stop. And since the universe is expanding as it gets older, you'll find that the expansion of the universe has carried your destination away from you while you were traveling toward it.

Because the expansion of the universe is accelerating, there are parts of the universe you won't be able to catch up to no matter *how* far you go. Current models of the expansion of the universe suggest this limit—known as the *cosmological event horizon*—is probably about a third of the way to the edge of the observable universe.

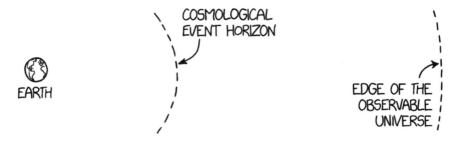

The Hubble space telescope has zoomed in on seemingly empty areas of the sky and taken photographs that show seas of dim, distant galaxies. Some of the larger, brighter galaxies in the photos are within our event horizon, so you could eventually reach them with your scooter, but most of them are beyond that limit. No matter how fast you accelerate toward them, the expansion of the universe will carry them ever farther away.

If you keep holding down the accelerator to chase these unreachable galaxies, they'll continue to grow more distant—but you'll find yourself plunging forward ever faster in time. After 30 years, the universe will be 10 trillion years old, and only the smallest and faintest long-lived stars will remain. After 40 years, even those stars will have burned out, and you'll find yourself in a dark, cold universe, lit only by intermittent flashes when the drifting husks of cold, dead stars happen to collide.

No matter how fast you go, you'll never get to the *edge* of the universe. But you can reach the *end*.

CHAPTER 27

How to Be on Time

There are two main ways to arrive somewhere more quickly: traveling faster, and leaving earlier.

OPTIONS

1. TRAVEL FASTER
2. LEAVE EARLIER

To learn how to travel faster, you can consult chapter 26: How to Get Somewhere Fast.

Leaving earlier is harder; it involves conscientiousness and realistic planning. To learn how to be better at those things, you should probably look for a different book.

If you rule out leaving earlier and traveling faster, then it seems like you're stuck. But you do have one more option: *alter the flow of time.*

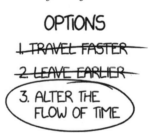

OPTIONS

1. ~~TRAVEL FASTER~~
2. ~~LEAVE EARLIER~~
3. ALTER THE FLOW OF TIME

This approach isn't necessarily as implausible as it sounds. When Einstein was studying the movement of electromagnetic waves through space, he was puzzled by how Maxwell's equations seemed to imply that an electromagnetic wave can never appear stationary relative to any observer. The equations suggested that you could never catch up to a light wave and see it frozen in place—no matter how fast you went, you'd measure the light moving past you at the same number of miles per hour. This led Einstein to realize that something must be wrong with our idea of "miles" and "hours," and his theories explained how time flows differently for different observers depending on how fast they're going.

Messing around with time earned Einstein fame, immortality, and a Nobel Prize,[1] so maybe it can get you where you're going on time. (And if not, maybe you'll get a Nobel Prize as a consolation.)

WE IN THE SWEDISH ROYAL FAMILY FELT BAD THAT YOU WERE LATE TO YOUR APPOINTMENT.

HERE'S A NOBEL PRIZE.

THANKS.

"Altering the flow of time" doesn't necessarily need to involve anything complicated; the simplest way to do it is to ask everyone to change their clocks. Many of us already do this twice a year thanks to daylight saving time. After all, clock time is a social construct. If you can get everyone to agree to turn their clocks back by an hour, then the time changes, potentially giving you an extra hour to get where you're going.

Time zones feel official and permanent, but they're more freewheeling than you might think. There's no international organization that has to approve time zone boundaries. Instead, every country has the authority to set their own clocks however they want, whenever they want. If the government of a country wakes up one morning and decides to move all their clocks back five hours, no one can stop them.

When a country messes with the flow of time without enough warning, it can cause some headaches. In March of 2016, the Cabinet of Ministers of Azerbaijan decided to cancel daylight saving time 10 days before it was scheduled to start. Software companies had to rush out updates, schedules had to be revised, and airlines had to decide whether flights should depart at the time on the ticket or an hour earlier. Heydar Aliyev International Airport just told everyone to arrive 3 hours early for their flights.

---

1   The Nobel Committee didn't actually give him the prize for the space-time stuff, in part because it was still considered revolutionary and not fully tested. Luckily, he published four papers in 1905, any *one* of which would probably be worthy of a Nobel, so they gave it to him for one of the more conventional ones.

Countries usually try to give more than 10 days of warning before changing their clocks, but they don't have to. In principle, if you're running late for an appointment, you can contact your government and ask them to move the clocks back.

HELLO, GOVERNMENT? I'M A CITIZEN RUNNING LATE FOR A MEETING—WHO DO I TALK TO ABOUT FIXING THAT?

In the United States, state legislatures *can* control whether daylight saving time is observed, but they can't control when it starts or stops. To gain an extra hour, you'll need to contact the federal government.

Federal law currently specifies 9 standard time zones and sets the time in each relative to Coordinated Universal Time, or UTC—from its French acronym—an international timekeeping system defined by the International Bureau of Weights and Measures. Congress can change this law, but you don't necessarily have to go through Congress to get your clock adjusted. By law, the secretary of transportation has the power to unilaterally move territory from one time zone to another. If you're in the mainland United States, you can potentially have the clock moved back by as much as eight hours simply by calling the Department of Transportation and asking nicely.

HELLO, DEPARTMENT OF TRANSPORTATION? BIG FAN OF YOUR WORK. I'M A LONGTIME SUPPORTER OF THINGS MOVING FROM PLACE TO PLACE.

LISTEN, I NEED TO ASK A FAVOR.

The secretary can't create new time zones, though. If you want to change the time to something other than the 9 standard values, you'll need to go through Congress. But if you can convince them to help you, you can set the time to whatever you want. In fact, you could – in principle – set the *date* to whatever you wanted. You could shift your house, town, or entire country forward by 24 hours... or backward by 65 million years.

SPRING FORWARD, FALL WAAAAY BACK.

In 2010, religious radio host Harold Camping predicted that the end of the world would begin with the Rapture on May 21, 2011, at 6:00 p.m. local time. Since the apocalypse happened according to *local* time, this meant that the end of the world was supposed to begin at the Republic of Kiribati in the Pacific Ocean, just west of the International Date Line, and sweep westward around the planet, time zone by time zone.

If some country wants to check whether the world ends at some future date, they can simply pass a law advancing their clock to, say, 12:00 p.m. on January 1, 3019, and then take a look around. If nothing happens, they can move the clocks back, and we'll all know that the next thousand years are safe – at least from apocalypses that happen in local time.

THE END IS NIGH!

DOOM

NO, IT'S COOL. BELGIUM CHANGED ITS CLOCKS FORWARD TO JANUARY 4099, AND EVERYTHING WAS STILL THERE.

If you're unable to persuade the government to change the clocks for you, or if your appointment is scheduled in UTC, you're stuck. You can't get more time for your appointment unless you can alter UTC itself.

ATOMIC CLOCKS

UTC is based on a network of precise atomic clocks. Atomic clocks measure the passage of time by precisely measuring the oscillation of cesium atoms using light. But we know, thanks to Einstein, that the passage of time isn't constant. In a strong gravitational field, light – and time itself – slows down. If you put a large spherical weight next to an atomic clock, the additional gravity will cause it to tick more slowly.

Unfortunately, you can't just tamper with *one* atomic clock. The International Bureau of Weights and Measures uses measurements from several hundred atomic clocks spread out around the world and averages them together to produce a single global time standard. If you wanted to artificially alter time, you'd have to slow down *all* those clocks together; if you just messed with one, they'd quickly notice the outlier.

Let's suppose you snuck into each atomic clock facility with a ball of lead 1 foot in diameter hidden in your backpack, and left it near the clock. (You'd have to be pretty strong, since that ball will weigh almost 400 pounds!)

If you managed to hide the ball right next to the atomic clock's timekeeping element, it would only slow the clock by about 1 part in 10^{24} – equivalent to about a hundred nanoseconds over the next 4 billion years.

A 200-meter-wide ball of lead would be only a little more effective, adding an extra nanosecond to the clocks every century or so. It would also be impossible to manufacture and move... and pretty difficult to hide.

If UTC is based on atomic clocks, and you can't mess with atomic clocks, then it seems like you can't mess with UTC. But UTC isn't *exactly* based on atomic clocks. It has an irregularity, one which could potentially help give you a little more time to get to your appointment... or, if you leave on time, make you show up too early.

CHANGING DAY LENGTHS

Our atomic clocks are more precise and regular than the Earth's spin. We used to define the length of a second in terms of the Earth's rotation, but a second whose length changes over time is inconvenient for physics, engineering, and timekeeping in general, so in 1967 the length of the second was officially and permanently frozen to match atomic clocks. A day is supposed to be 24 hours, or 86,400 seconds, but as of the late 2010s the Earth takes about 86,400.001 seconds on average to make one full turn relative to the Sun. In other words, the Earth is a millisecond too slow. That extra millisecond each day gradually adds up. After about a thousand days a perfect clock would drift 1 full second out of sync with the Sun.

The day may only be a few milliseconds too long *now*, but it isn't going to stay that way. Thanks to the Moon, the Earth's rotation is slowing down.

The Moon's gravity pulls harder on the nearer parts of the Earth than the farther parts. As the Earth spins, the water (and, to a lesser extent, the land) sloshes around slightly to adjust to the shifting force, which we experience as tides. The Earth spins faster than the Moon orbits, and the gravitational pull between those sloshing oceans and the Moon create a very slight gravitational "drag" between the two bodies. This has the effect of pulling the Moon forward – flinging it into a wider orbit – while also slowing down the Earth.[2]

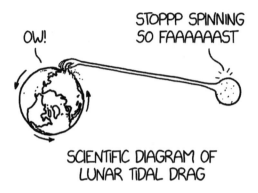

OW!

STOPPP SPINNING
SO FAAAAAAST

SCIENTIFIC DIAGRAM OF
LUNAR TIDAL DRAG

LEAP SECONDS

UTC has no time zones or daylight saving, but it's adjusted every so often – very slightly – to keep the clocks in sync with the Earth's rotation. These changes take the form of leap seconds.

Leap seconds are added by the International Earth Rotation and Reference Systems Service, which carefully tracks the Earth's rotation and decides when a new leap second is needed. The leap second is added right before midnight on the last day of a month, usually June or December. The second is inserted between 11:59:59 p.m. and 12:00:00 a.m., and represented as 11:59:60 p.m.

11:59:59 PM

11:59:60 PM

12:00:00 AM

---

2   At least, it *should* be slowing down. Over the long term, the Earth's spin has been slowing steadily, but in recent decades it's actually sped up a little. Since about 1972 – coincidentally, when we started adding leap seconds – the time it takes the Earth to complete one spin has gotten a few milliseconds shorter. This is likely due to impossible-to-predict turbulence in the currents in the Earth's molten outer core, but no one really knows for sure. It's not too unusual – the Earth has sped up and slowed down a number of times over the last few centuries – and it's unlikely to continue for too much longer. But it's still a little strange to think that the Earth is speeding up and no one knows why.

When a leap second is inserted, any event scheduled after that date is pushed back by 1 second. If your appointment is more than a month or two in the future, you could potentially gain extra seconds by convincing the International Earth Rotation and Reference Systems Service that leap seconds are needed.

To get *more* leap seconds, you need to slow the Earth down faster.

Whenever mass shifts from the equator to the poles, Earth speeds up. The daily motion of air between the poles and the Equator causes Earth's speed to wobble up and down, and over longer periods, redistribution of mass due to climate cycles, melting ice sheets, and post-glacial rebound all have their own effects.

That means if you live in a tropical or temperate zone, you can speed up the Earth simply by walking to one of the poles, and you can slow it down by walking from the pole back to the equator.

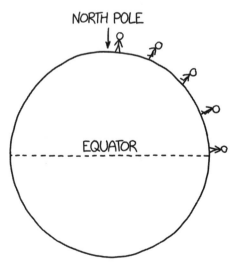

The effect won't be very large. A single person moving from the pole to the equator will lengthen the day by less than 1 part in 10,000,000,000,000,000,000,000. It would take a million years for that discrepancy to add up to a single extra nanosecond. If you want to gain 1 extra leap second within the next year, you'll need to move *60 trillion tons* of material from the poles to the Equator.

Even if you use something as dense as gold, that's over 3,000 cubic kilometers – enough to wrap the equator in a mile-high wall 150 feet thick. That's clearly impossible...

UNLESS...

... unless you can find a magical entity at the North Pole that can produce an unlimited volume of expensive objects, and magically transport them with impossible efficiency from the pole all around the world.

CHAPTER 28

How to Dispose of This Book

If you're done with this book and decide you want to get rid of it, the simplest thing to do is to give it to someone else.

But maybe you don't want to give it away. Perhaps you've written notes in the margin that you don't want anyone to see. Perhaps you simply didn't like it. Or perhaps you plan to use the information in the book as part of some kind of supervillain-style plot, and you're trying to buy up every copy and destroy them so no one else can use it to gain an advantage over you.[1]

If, for whatever reason, you do decide to permanently dispose of this or any other book, here are a few tips.

---

1 *Editor's note: If you would like to purchase all existing copies of this book, please contact the Riverhead sales department.*

AIR DISPOSAL

In a pinch, this book can double as a source of power. The pages contain about 8 megajoules of chemical energy, energy that was originally collected from the Sun by leaves.

Plants are made of air. The carbon in wood comes from CO_2 collected from the air, which is combined with water (H_2O) through photosynthesis. This book is made from air, water, and sunlight. If the pages are incinerated, the carbon turns back into CO_2 and water, releasing the captured sunlight. When wood, oil, or paper burns, the heat of the fire is the heat from that sunlight.

Eight megajoules is roughly equal to the energy in a cup of gasoline. If your car's fuel economy is 30 miles per gallon on the highway when driving at 55 mph, and you convert it to run on copies of this book instead of gas, it will burn through 30,000 words per minute, several dozen times faster than the word consumption rate of a typical human.

$$55 \text{ mph} \times \frac{65,000 \frac{\text{words}}{\text{book}}}{30 \frac{\text{miles}}{\text{gallon}} \times 1 \frac{\text{cup}}{\text{book}}} = 30,000 \frac{\text{words}}{\text{minute}}$$

THIS ENGINE HAS THE POWER OUTPUT OF 200 HORSES AND THE BOOK CONSUMPTION OF DOZENS OF LIBRARIANS.

OCEAN DISPOSAL

The carbon in a book can also be mixed into water. If the book is incinerated, its carbon and hydrogen will be converted to CO_2 and water. The water vapor will fall as rain and likely end up in the ocean: Half of the CO_2 released into the atmosphere by combustion will also end up absorbed by the ocean, forming several septillion molecules of carbonic acid. If it were mixed evenly into the air and ocean, each cup of seawater and each lungful of air would contain several thousand molecules from the book.

TIME DISPOSAL

If you set this book down on the ground and walked away, and no one ever touched it again, what would happen to it?

Depending on the climate in your area, it may not last that long. Humans can't eat paper, but the energy stored in cellulose – the same energy released when you burn it – is appetizing to a wide variety of microorganisms. These organisms need warmth and high levels of moisture to flourish, so books are usually safe on your shelf indoors. If you abandon it in a cool, dry cave or shady spot in the desert, it might last for centuries. But once the book gets wet on a warm day, organisms – generally fungi – will start to devour the cellulose. The pages will be digested and eventually mixed into the environment.

If the book is protected from decomposition, its fate may depend on local geology. If you left it in an area where sediment is being deposited, such as a low-lying flood plain, it will gradually be buried. If it's in an area where sediment is being *eroded*, like a rocky mountainside, it will almost certainly be broken down and carried away by the wind and water. Rock erodes at rates measured in fractions of a millimeter per year, so if this book were made of rock, it would probably take centuries or millennia to erode away. Since paper is much softer than rock, it probably won't take anywhere near that long. The paper will weather away and disintegrate, and the information printed on it will be lost.

DISPOSAL OF AN INDESTRUCTIBLE OR CURSED BOOK

It's technically possible that the copy of this book that you're reading is indestructible. Sure, it seems unlikely, but you can't definitively rule it out without trying. There's no nondestructive test for indestructibility.

If you ever get your hands on a book that you want to get rid of but can't destroy—either because the paper is too strong, or because of some kind of Hogwarts library/Ring of Power/Jumanji situation—what should you do? Where do you put something if you want to get rid of it forever?

We face this problem with nuclear waste. We want to get rid of the stuff, but there's no way to destroy it or convert it into a less-dangerous form, because incinerating or vaporizing radioactive waste doesn't reduce the radioactivity. With enough heat, you can destroy anything by breaking its molecules apart into their constituent atoms. But doing this to radioactive waste doesn't help—because the atoms *themselves* are the problem.

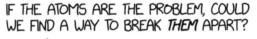

IF THE ATOMS ARE THE PROBLEM, COULD WE FIND A WAY TO BREAK *THEM* APART?

LISTEN, THAT'S EXACTLY HOW WE GOT INTO THIS MESS IN THE FIRST PLACE.

Since we can't destroy radioactive waste, we generally try to put it somewhere where it won't bother us. Collecting it all in a single location makes sense—there's not *that* much waste, volume-wise—so we could just pick a spot, put all of our waste there, and then seal it off as permanently as possible, monitoring the site indefinitely, with some kind of warning signs to steer future civilizations away from digging it up.[2]

Currently, America's only long-term permanent underground waste disposal site is a series of chambers 2,000 feet below the New Mexico desert. The complex, called the Waste Isolation Pilot Plant, continues to accept a portion of our nuclear waste, but until a new permanent disposal site is chosen or the WIPP facility is expanded, we're solving this problem the way we so often do: by trying not to think about it and hoping it goes away.

WASTE ISOLATION PILOT PLANT

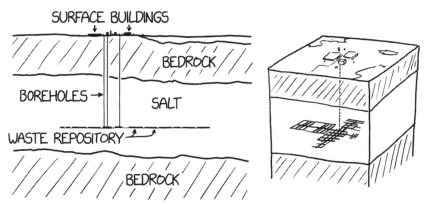

The WIPP tunnels in New Mexico are dug through an ancient layer of rock salt half a kilometer thick. Salt tunnels are particularly convenient for waste disposal because the salt "flows" very slowly. If you dig a tunnel through salt and then abandon it, the tunnel will gradually contract and seal itself off.

SALT CREEP

---

2   In the 1990s, a panel of experts was assembled to ponder the question of how to create markings that will make it clear to future civilizations that they shouldn't dig up our nuclear waste, involving informational inscriptions in various languages, diagrams, and ominous sculptures. The whole exercise is a strange combination of gloom and optimism – gloom that we've created something so dangerous that it poses a threat not only to us, but to future civilizations – and optimism that there will *be* future civilizations, long after we've been forgotten, who might read and understand the messages we leave for them.

To dispose of this book at the WIPP facility, you could dig a cavity off to one side of a tunnel[3] and leave this book inside. After a few decades, the cavity will close up, entombing the text in salt.

There's another idea for how to get rid of our radioactive waste which, according to proponents, could be both cheaper and safer than a WIPP-style facility: dropping it into very deep boreholes.

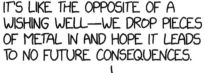

IT'S LIKE THE OPPOSITE OF A WISHING WELL—WE DROP PIECES OF METAL IN AND HOPE IT LEADS TO NO FUTURE CONSEQUENCES.

The WIPP facility is about half a kilometer deep, but boreholes for oil drilling and geological research[4] go much deeper. Some reach as far as 10 kilometers below the surface, down through the surface layers and deep into the underlying mass of ancient rock that makes up the core of the continent—what geologists refer to as the *crystalline basement*.[5]

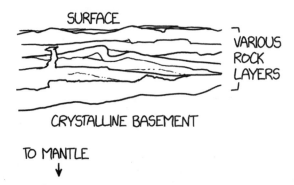

In many parts of the world, the rock in the crystalline basement has been isolated from the surface for billions of years. To dispose of something there, we could dig a long borehole straight down, drop the waste in, and then seal up the hole with layers of cement and expanding clay.

---

3   See chapter 3: How to Dig a Hole.

4   Mostly to look for oil.

5   If you asked me what the term *crystalline basement* meant before I learned it, my guesses would include: "Mario Kart level," "electronic music subgenre," "home improvement project," and "illegal synthetic drug."

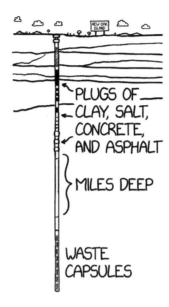

SUBDUCTION

Oceanic crust is recycled into the Earth's mantle through *subduction*, so people sometimes suggest putting our nuclear waste in an ocean trench and letting the Earth dispose of it for us. Unfortunately, subduction is pretty slow. If we lodge our waste a kilometer deep in a subducting plate, then wait 10,000 years...

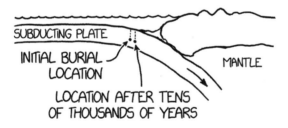

... it will have moved about 300 meters sideways.

SHOOT IT INTO THE SUN

People often suggest shooting our nuclear waste into the Sun, where it will be disintegrated and will either be carried away by the solar wind or sink to the Sun's core. The biggest problem with this idea is that rocket launches fail sometimes. If you send up 100 rockets full of many tons of radioactive debris, the odds are pretty good that one of the launches will fail—and it would be hard to come up with a *worse* thing to do with nuclear waste than packing it into a rocket and blowing it up high in the atmosphere.

However, if you're disposing of a single cursed or indestructible book, then the Sun seems more attractive as a disposal site. A book only requires a single launch, which reduces the risk of failure, and if the book is indestructible, then if the launch fails, you just need to recover it and try again.

A tip for dropping things into the Sun: launching directly to the Sun from Earth is really difficult—it actually takes more fuel than launching something out of the Solar System completely. A more efficient way to reach the Sun is to launch something to the far outer Solar System—possibly with the help of gravity assists from the planets. When it's far from the Sun, it will be moving very slowly, and it will only take a little extra fuel to slow it to a halt—after which it will fall directly toward the Sun. It takes much longer than a direct launch, but only requires a fraction of the fuel.

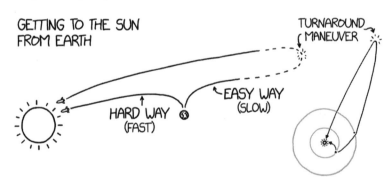

But perhaps you don't *want* to destroy this book. Perhaps you want to preserve it.

HOW TO PRESERVE THIS BOOK

Leaving this book in a borehole or salt mine could in theory preserve it for millions or perhaps billions of years, if it is not disturbed by tectonic activity, meddling humans, or hungry microbes. But to *really* preserve a book, you may want to remove it from Earth completely.

ESA's Rosetta spacecraft and Philae lander reached comet 67P/Churyumov-Gerasimenko in 2014. The spacecraft carried a nickel-titanium disc etched with 6,000 pages of text in 1,000 different human languages. This disc, constructed by the Long Now Foundation, is designed to last for millennia. The comet will likely remain in a stable orbit for millions of years, so if the disc is in a sheltered location on the surface of the comet, protected from micrometeorites and cosmic rays, it will probably remain intact and readable for longer than even the longest-lived civilization.

Written words are a message to the future. The person reading it is always further ahead in time than the person writing it. I don't know what date it is when you're reading these words, where you are, or what you're trying to do. But wherever you are and whatever problems you're trying to solve, I hope this book has helped. There's a giant, weird world out there. Ideas that sound good can have terrible consequences, and ideas that sound ridiculous can turn out to be revolutionary. Sometimes you can figure out which ones work ahead of time, and sometimes you just have to try them and see what happens.

(But you might want to stand at a safe distance.)

Acknowledgements

A lot of people helped make this book possible.

Many people lent me their expertise and resources. Thank you to Serena Williams and Alexis Ohanian for their willingness to sacrifice a drone for science, and to Kate Darling for telling us that it was probably ok to do so. Thank you to Col. Chris Hadfield for answering the most ridiculous questions I could think of, and to Katie Mack for warning me not to end the universe. And thank you to Christopher Night and Nick Murdoch for their help with equations and measurements.

Thank you to Kathleen Weldon and the Roper Center staff for digging up strange polling data, and to *HuffPost* polling editor Ariel Edwards-Levy for answering my questions about public opinion. Thank you to Anna Romanov and David Allen for making their undergraduate project available, and to Dr. Reuben Thomas for sharing his research on friendship. Thank you to Greg Leppert for helping to arrange the Infrasonata, and thank you to the ants that got into Waldo Jaquith's house, which led him to ask me for help building a lava moat.

Thank you to Christina Gleason for molding my text and drawings into the shape of a book, and providing wise and invaluable advice throughout. Thank you to Derek for helping to make this whole thing happen, and thank you to Seth Fishman, Rebecca Gardner, Will Roberts, and the rest of the team at Gernert.

Thank you to my positively heroic editor Courtney Young, and to the rest of the team at Riverhead, including Kevin Murphy, Helen Yentus, Annie Gottleib, Ashley Garland, May-Zhee Lim, Jynne Martin, Melissa Solis, Caitlin Noonan, Gabriel Levinson, Linda Friedner, Grace Han, Claire Vaccaro, Taylor Grant, Mary Stone, Nora Alice Demick, Kate Stark, and publisher Geoff Kloske.

And thank you to my wife, for teaching me half the stuff in this book and exploring this big, weird, and exciting world with me.

References

1. How to Jump Really High

Carter, Elizabeth J., E. H. Teets, and S. N. Goates, "The Perlan Project: New Zealand flights, meteorological support and modeling," in *Proc. 19th Int. Cont. on IIPS, 83rd AMS Annual Meeting*, no. 1.2 (2003).

Hirt, Christian et al., "New Ultrahigh-Resolution Picture of Earth's Gravity Field," *Geophysical Research Letters* 40, no. 16 (August 2013): 4279-4283.

Teets, Edward H., Jr., "Atmospheric Conditions of Stratospheric Mountain Waves: Soaring the Perlan Aircraft to 30 km," in *10th Conference on Aviation, Range, and Aerospace Meteorology* (2002).

2. How to Throw a Pool Party

Arctic Monitoring and Assessment Programme, *Snow, Water, Ice and Permafrost in the Arctic (SWIPA) 2017* (Oslo 2017).

Trenberth, Kevin E. and Lesley Smith, "The Mass of the Atmosphere: A Constraint on Global Analyses," *Journal of Climate* 18, no. 6 (March 2005): 864-875.

Wellerstein, Alex, "Beer and the Apocalypse," *Restricted Data*, September 5, 2012, http://blog.nuclearsecrecy.com/2012/09/05/beer-and-the-apocalypse/.

3. How to Dig a Hole

Nevola, V. René, "Common Military Task: Digging," in *Optimizing Operational Physical Fitness* (RTO/NATO, 2009), 4-1-68.

United States Department of Labor, "Occupational Employment and Wages, May 2017," Bureau of Labor Statistics, last modified March 30, 2018, https://www.bls.gov/oes/current/oes472061.htm.

4. How to Play the Piano

Katharine B. Payne, William R. Langbauer Jr., Elizabeth M. Thomas, "Infrasonic Calls of the Asian Elephant (Elephas Maximus)," *Behavioral Ecology and Sociobiology* 18, no. 4 (February 1986): 297-301.

6. How to Cross a River

Buffalo Morning Express, February 10, 1848.

Glauber, Bill, "On Solid or Liquid, Give It the Gas," *Journal Sentinel*, July 18, 2009, http://archive.jsonline.com/news/wisconsin/51105382.html/.

Historic Lewiston, *Lewiston History Mysteries*, Summer 2016, http://historiclewiston.org/wp-content/uploads/2016/08/Homan-Walsh-Falls-Kite-3.pdf.

"Incidents at the Falls," *Buffalo Commercial Advertiser*, July 13, 1848.

"Niagara Suspension Bridge," *Buffalo Daily Courier*, February 3, 1848.

Perkins, Frank C., "Man-Carrying Kites in Wireless Service," *Electrician and Mechanic* 24 (January-June 1912): 59.

Robinson, M., "The Kite that Bridged a River," 2005, http://kitehistory.com/Miscellaneous/Homan_Walsh.htm.

7. How to Move

Federal Emergency Management Agency, "Appendix C, Sample Design Calculations" in *Engineering Principles and Practices for Retrofitting Flood-Prone Residential Structures* (FEMA 2009), C–1-37

Piasecki Aircraft Corporation, "Multi-Helicopter Heavy Lift System Feasibility Study" (Naval Air Systems Command, 1972).

8. How to Keep Your House from Moving

AK Stat. § 09.45.800 (Alaska 2017).

California Code of Civil Procedure, chapter 3.6, Cullen Earthquake Act, § 751.50 (1972)).

Joannou v. City of Rancho Palos Verdes, B241035 (CA Ct. App. 2013).

Offord, Simon, "Court Denies Request to Adjust Lot Lines After Landslide," Bay Area Real Estate Law Blog, accessed March 28, 2019, https://bayarearealestatelawyers.com/real-estate-law/court-denies-request-to-adjust-lot-lines-after-landslide.

Pallamary, Michael J. and Curtis M. Brown, "Land Movements and Boundaries" from *The Curt Brown Chronicles, The American Surveyor* 10, no. 10 (2013): 49-50.

Schultz, Sandra S. and Robert E. Wallace, "The San Andreas Fault," U.S. Geological Survey, last modified November 30, 2016, https://pubs.usgs.gov/gip/earthq3/safaultgip.html.

Theriault v. Murray, 588 A.2d 720 (Maine 1991).

White, C. Albert, "Land Slide Report" (Bureau of Land Management, 1998), https://www.blm.gov/or/gis/geoscience/files/landslide.pdf.

9. How to Build a Lava Moat

Heus, Ronald and Emiel A. Denhartog, "Maximum Allowable Exposure to Different Heat Radiation Levels in Three Types of Heat Protective Clothing," *Industrial Health* 55, no. 6 (November 2017): 529-536.

Keszthelyi, Laszlo, Andrew J. L. Harris, and Jonathan Dehn, "Observations of the Effect of Wind on the Cooling of Active Lava Flows," *Geophysical Research Letters* 30, no. 19 (October 2003): 4-1-4.

Torvi, D. A., G. V. Hadjisophocleous, and J. K. Hum, "A New Method for Estimating the Effects of Thermal Radiation from Fires on Building Occupants,"

Proceedings of the ASME Heat Transfer Division (National Research Council of Canada, 2000): 65-72.

"What Is Lava Made Of?," *Volcano World*, Oregon State University, http://volcano.oregonstate.edu/what-lava-made.

Wright, Thomas L., "Chemistry of Kilauea and Mauna Loa Lava in Space and Time" (U.S. Geological Survey 1971), https://pubs.usgs.gov/pp/0735/report.pdf.

10. How to Throw Things

Cronin, Brian, "Did Walter Johnson Accomplish a Famous George Washington Myth?," *Los Angeles Times,* September 21, 2012, https://www.latimes.com/sports/la-xpm-2012-sep-21-la-sp-sn-walter-johnson-george-washington-20120921-story.html.

McLean, Charles, "Johnson Twice Throws a Dollar Across the Turbid Rappahannock," *New York Times*, February 23, 1936.

Ragland, K. W., M. A. Mason, and W. W. Simmons, "Effect of Tumbling and Burning on the Drag of Bluff Objects," *Journal of Fluids Engineering* 105, no. 2 (June 1983): 174-178.

Sprague, Robert et al., "Force-Velocity and Power-Velocity Relationships during Maximal Short-Term Rowing Ergometry," *Medicine & Science in Sports & Exercise* 39, no. 2 (February 2007): 358-364.

Taylor, Lloyd W., "The Laws of Motion Under Constant Power," *The Ohio Journal of Science* 30, no. 4 (July 1930): 218-220.

11. How to Play Football

Goff, John Eric, "Heuristic Model of Air Drag on a Sphere," *Physics Education* 39, no. 6 (November 2004): 496-499.

White, Frank M., *Fluid Mechanics* (New York: McGraw Hill, 2016).

12. How to Predict the Weather

"Daniel K. Inouye International Airport, Hawaii," Weather Underground, July 2017, https://www.wunderground.com/history/monthly/us/hi/honolulu/PHNL/date/2017-7.

Gough, W. A., "Theoretical Considerations of Day-to-Day Temperature Variability Applied to Toronto and Calgary, Canada Data," *Theoretical and Applied Climatology* 94, no. 1-2 (September 2008): 97-105.

"Honolulu, HI, NOAA Online Weather Data," National Weather Service Forecast Office, accessed May 3, 2019, https://w2.weather.gov/climate/xmacis.php?wfo=hnl.

Roehrig, Romain, Dominique Bouniol, Francoise Guichard, Frédéric Hourdin, and Jean-Luc Redelsperger, "The Present and Future of the West African Monsoon," *Journal of Climate* 26 (September 2013): 6471-6505.

Thompson, Philip, "Philip Thompson Interview," interview by William Aspray, Charles Babbage Institute, University of Minnesota, December 5, 1986, transcript.

Trenberth, Kevin E., "Persistence of Daily Geopotential Heights over the Southern Hemisphere," *Monthly Weather Review* 113 (January 1985): 38-53.

13. How to Play Tag

Bethea, Charles, "How Fast Could Usain Bold Run the Mile," *The New Yorker*, August 1, 2016, https://www.newyorker.com/sports/sporting-scene/how-fast-would-usain-bolt-run-the-mile.

Dawson, Andrew, "Belgian Dentist Breaks Appalachian Trail Speed Record," *Runner's World,* August 29, 2018, https://www.runnersworld.com/news/a22865359/karel-sabbe-breaks-appalachian-trail-speed-record/.

Krzywinski, Martin, "The Google Maps Challenge—Longest Google Maps Driving Routes," *Martin Krzywinski Science Art*, last modified June 13, 2017, http://mkweb.bcgsc.ca/googlemapschallenge/.

Krzywinski, Martin, "Longest possible Google Maps route?," xkcd forum, January 30, 2012, http://forums.xkcd.com/viewtopic.php?f=2&t=65793&p=2872419#p28724199.

"Thru-Hiking," Appalachian Trail Conservancy, accessed March 28, 2019, http://www.appalachiantrail.org/home/explore-the-trail/thru-hiking.

14. How to Ski

"Facts on Snowmaking," National Ski Areas Association, accessed March 28, 2019, https://www.nsaa.org/media/248986/snowmaking.pdf.

Friedland, Lois, "Tanks for the Snow," *Ski,* March 1988, 13.

Louden, Patrick B. and J. Daniel Gezelter, "Friction at Ice-Ih/Water Interfaces Is Governed by Solid/Liquid Hydrogen-Bonding," *The Journal of Physical Chemistry* 121, no. 48 (November 2017): 26764-26776.

"Polarsnow," Polar Europe, accessed Marche 28, 2019, https://polareurope.com/polar-snow/.

Rosenberg, Bob, "Why is Ice Slippery?," *Physics Today* 58, no. 12 (December 2005): 50.

Scanlan, Dave from "Like It or Not, Snowmaking is the Future," interview by Julie Brown, *Powder*, August 29, 2017, https://www.powder.com/stories/news/like-not-snowmaking-future/.

15. How to Mail a Package

"Apollo 13 Press Kit," NASA, April 2, 1970, https://www.hq.nasa.gov/alsj/a13/A13_PressKit.pdf.

Atchison, Justin Allen, "Length Scaling in Spacecraft Dynamics" (PhD diss., Cornell University, 2010).

The Corona Story, National Reconnaissance Office, November 1987 (Partially declassified and released under the Freedom of Information Act (FOIA), June 30, 2010).

Janovsky, R. et al., "End-of-life De-orbiting Strategies for Satellites," paper presented at Deutscher Luft- und Raumfahrtkongress, Stuttgart, Germany, September 2002.

Peck, Mason, "Sometimes Even a Low Ballistic Coefficient Needs a Little Help," *Spacecraft Lab*, May 5, 2014, https://spacecraftlab.wordpress.com/2014/05/05/sometimes-even-a-low-ballistic-coefficient-needs-a-little-help/.

Portree, David S. F. and Joseph P. Loftus, Jr., *Orbital Debris* (Houston: NASA, 1999).

Singer, Mark, "Risky Business," *The New Yorker*, July 14, 2014, https://www.newyorker.com /magazine/2014/07/21/risky-business-2.

"Taco Bell Cashes In on Mir," BBC News, March 20, 2001, http://news.bbc.co.uk/2/hi/americas/1231447.stm.

Yamaguchi, Mari, "Can an Origami Space Shuttle Fly from Space to Earth," *USA Today*, March 27, 2008, https://usatoday30.usatoday.com/tech/science /space/2008-03-27-origami-space-shuttle_N.htm/.

16. How to Power your House (Earth)

"Appendix A: Frequently Asked Questions" in *Woody Biomass Desk Guide and Toolkit* adapted by Sarah Ashton, Lauren McDonnell, and Kiley Barnes (Washington, D.C.: National Association of Conservation Districts): 119-130.

Arevalo, Ricardo, Jr., William F. McDonough, and Mario Luong, "The K/U Ration of the Silicate Earth," *The Earth and Planetary Science Letters* 278, no. 3-4 (February 2009): 361-369.

Chacón, Felipe, "The Incredible Shrinking Yard!," Trulia, October 18, 2017, https://www.trulia.com/research /lot-usage/.

"Environmental Impacts of Geothermal Energy," Union of Concerned Scientists, accessed March 28, 2019, https://www.ucsusa.org/clean_energy/our-energy -choices/renewable-energy/environmental-impacts -geothermal-energy.html.

"Coal Explained: How Much Coal is Left," U.S. Energy Information Administration, last modified November 15, 2018, https://www.eia.gov/energyexplained/index .php?page=coal_reserves.

"How Much Do Solar Panels Cost for the Average House in the US in 2019?," SolarReviews, last modified March 2019, https://www.solarreviews.com/solar-panels /solar-panel-cost/.

"How Much Electricity Does an American Home Use?," Frequently Asked Questions, U.S. Energy Information Administration, last modified October 26, 2018, https://www.eia.gov/tools/faqs/faq.php?id=97&t=3.

NOAA National Centers for Environmental Information, "Climate at a Glance: National Time Series," accessed March 28, 2019, https://www.ncdc.noaa.gov/cag/.

Rinehart, Lee, "Switchgrass as a Bioenergy Crop," ATTRA (NCAT, 2006).

"Section 6: Geography and Environment" in *Statistical Abstract of the United States: 2004-2005* (U.S. Census Bureau, 2006), 211-236.

"Solar Maps," National Renewable Energy Laboratory, accessed March 28, 2019, https://www.nrel.gov/gis /solar.html.

"Solar Resource Data and Tools," National Renewable Energy Laboratory, accessed March 28, 2019, https ://www.nrel.gov/grid/solar-resource/renewable -resource-data.html.

"Transparent Cost Database," Open Energy Information, last modified November 2015, https://openei.org /apps/TCDB/transparent_cost_database#blank.

" U.S. Crude Oil and Natural Gas Proved Reserve, Year-End 2017," U.S. Energy Information Administration, last modified November 29, 2018, https://www.eia .gov/naturalgas/crudeoilreserves/.

"U.S. Uranium Reserves Estimates,"U.S. Energy Information Administration, last modified July 2010, https://www.eia.gov/uranium/reserves/.

17. How to Power your House (on Mars)

Boardman, Warren P. et al., Firestream Ram Air Turbine, U.S. Patent 2,986,219 filed May 27, 1977, issued May 30, 1961.

"Country Comparison: Electricity—Consumption," *The World Factbook* (Washington, DC: Central Intelligence Agency), last modified 2016, https://www.cia.gov /library/publications/resources/the-world-factbook /fields/253rank.html.

Hoffman, N., "Modern Geothermal Gradients on Mars and Implications for Subsurface Liquids," Conference on the Geophysical Detection of Subsurface Water on Mars (August 2001).

Hollister, David, "How Wolfe's Tether Spreadsheet Works," *Hop's Blog*, December 16, 2015, http ://hopsblog-hop.blogspot.com/2015/12/how-wolfes -tether-spreadsheet-works.html.

"Sounds on Mars," The Planetary Society, accessed March 29, 2019, http://www.planetary.org/explore/projects /microphones/sounds-on-mars.html.

Weinstein, Leonard M., "Space Colonization Using Space-Elevators from Phobos," AIP Conference Proceedings (American Institute of Physics, 2003): 1227-1235.

18. How to Make Friends

Gallup Organization, Gallup Poll (AIPO), January 1990, USGALLUP.922002.Q20, Cornell University, Ithaca, NY: Roper Center for Public Opinion Research, iPOLL.

National Institute for Transforming India, "Population Density (Per Sq. Km.)," last modified March 30, 2018, http://niti.gov.in/content/population-density-sq-km.

Thomas, Reuben J., "Sources of Friendship and Structurally Induces Homophily across the Life Course," *Sociological Perspectives* (February 11, 2019).

19. How to Send a File

Cisco, "Cisco Global Cloud Index: Forecast and Methodology, 2016-2021 White Paper," November 19, 2018, https://www.cisco.com/c/en/us/solutions /collateral/service-provider/global-cloud-index-gci /white-paper-c11-738085.html.

Erlich, Yaniv and Dina Zielinski, "DNA Fountain Enables a Robust and Efficient Storage Architecture," *Science* 355, no. 6328 (March 2017): 950-954.

Gibo, David L. and Megan J. Pallett, "Soaring Flight of Monarch Butterflies *Danaus Plexippus* (Lepidoptera: Danaidae), During the Late Summer Migration in Southern Ontario," *Canadian Journal of Zoology* 57, no. 7 (1979): 1393-1401.

"Intel/Micron 64L 3D NAND Analysis," *TechInsights*, accessed March 29, 2019, https://techinsights.com /technology-intelligence/overview/latest-reports /intel-micron-64l-3d-nand-analysis/.

Mizejewski, David, "How the Monarch Butterfly Population is Measured," National Wildlife Federation, February 7, 2019, https://blog.nwf.org/2019/02 /how-the-monarch-butterfly-population-is-measured/.

Morris, Gail, Karen Oberhauser, and Lincoln Brower, "Estimating the Number of Overwintering Monarchs in Mexico," Monarch Joint Venture, December 6, 2017, https://monarchjointventure.org/news-events/news /estimating-the-number-of-overwintering-monarchs -in-mexico.

Stefanescu, Constantí et al., "Long-Distance Autumn Migration Across the Sahara by Painted Lady Butterflies: Exploiting Resource Pulses in the Tropical Svannah," *Biology Letters* 12, no. 10 (October 2016).

Talavera, Gerard and Roger Vila, "Discovery of Mass Migration and Breeding of the Painted Lady Butterfly *Vanessa Cardui* in the Sub-Sahara," *Biological Journal of the Linnean Society* 120, no. 2 (February 2017): 274-285.

Walker, Thomas J. and Susan A. Wineriter, "Marking Techniques for Recognizing Individual Insects," *The Florida Entomologist* 64, no. 1 (March 1981): 18-29.

20. How to Charge Your Phone

Jacobson, Mark Z. and Cristina L. Archer, "Saturation Wind Power Potential and its Implications for Wind Energy," *Proceedings of the National Academy of Sciences of the United States of America* 109, no. 39 (September 2012): 15679-15684.

Max Planck Institute for Biogeochemistry, "Gone with the Wind: Why the Fast Jet Stream Winds Cannot Contribute Much Renewable Energy After All," ScienceDaily, November 30, 2011, https://www .sciencedaily.com/releases/2011/11/111130100013.htm.

Rancourt, David, Ahmadreza Tabesh, and Luc G. Fréchette, "Evaluation of Centimeter-Scale Micro Wind Mills," paper presented at *7th International Workshop on Micro and Nanotechnology for Power Generation and Energy Conversion App's,* Freiburg, Germany, November 2007.

Romanov, Anna Macquarie and David Allen, "A Bicycle with Flower-Shaped Wheels," Differential Geometry Final Project, Colorado State University, 2011.

World Energy Resources (London: World Energy Council, 2016).

21. How to Take a Selfie

Chang, Hsiang-Kuang, Chih-Yuan Liu, and Kuan-Ting Chen, "Search for Serendipitous Trans-Neptunian Object Occultation in X-rays," *Monthly Notices of the Royal Astronomical Society* 429, no. 2 (February 2013): 1626-1632.

Colas, F. et al., "Shape and Size of (90) Antiope Derived From an Exceptional Stellar Occultation on July 19, 2011," paper presented at *American Geophysical Union, Fall Meeting,* December 2011.

Larson, Adam M. and Lester Loschky, "The Contributions of Central versus Peripheral Vision to Scene Gist Recognition," *Journal of Vision* 9, no. 10 (September 2009): 6.1-16.

22. How to Catch a Drone

"All-Star Skills Competition 2012: Canadian Tire NHL Accuracy Shooting," Canadian Broadcasting Corporation, accessed March 29, 2019, https://www .cbc.ca/sports-content/hockey/nhlallstargame/skills /accuracy-shooting.html.

"Distance from Center of Fairway," PGA Tour, continuously updated, https://www.pgatour.com/stats /stat.02421.html.

Kawamura, Katsue et al., "Baseball Pitching Accuracy: An Examination of Various Parameters When Evaluating Pitch Locations," *Sports Biomechanics* 16, no. 3 (August 2017): 399-410.

Kempf, Christopher, "Stats Analysis: Running for Cover," Professional Darts Corporation, October 1, 2019, https://www.pdc.tv/news/2019/01/10 /stats-analysis-running-cover.

Landlinger, Johannes et al., "Differences in Ball Speed and Accuracy of Tennis Groundstrokes Between Elite and High-Performance Players," *European Journal of Sport Science* 12, no. 4 (October 2011): 301-308.

Michaud-Paquette, Yannick et al., "Whole-Body Predictors of Wrist and Shot Accuracy in Ice Hockey," *Sports Biomechanics* 10, no. 1 (March 2011): 12-21.

Morris, Benjamin, "Kickers Are Forever," *FiveThirtyEight*, January 28, 2015, https://fivethirtyeight.com/features /kickers-are-forever/.

Wells, Chris, "Stat Sheet: 10 Facts from Rio 2016 Olympics Entry List," World Archery, July 18, 2016, https://worldarchery.org/news/142029 /stat-sheet-10-facts-rio-2016-olympics-entry-list.

23. How to Tell if You're a Nineties Kid

"Figure 6. Yield of Atmospheric Nuclear Tests Per Year Shown by Bars," graph, from "Is There an Isotopic Signature of the Anthropocene?," *The Anthropocene Review* 1, no. 3 (December 2014): 8.

Goldman, G.S. and P.G. King, "Review of the United States Universal Vaccination Program: Herpes Zoster Incidence Rates, Cost-Effectiveness, and Vaccine Efficacy Based Primarily on the Antelope Valley Varicella Active Surveillance Project Data," *Vaccine* 31, no. 13 (March 2013): 1680-1694.

Gulson, Brian L. and Barrie R. Gillings, "Lead Exchange in Teeth and Bone—A Pilot Study Using Stable Lead Isotopes," *Environmental Health Perspectives* 105, no. 8 (August 1997): 820-824.

Gulson, Brian L., "Tooth Analyses of Sources and Intensity of Lead Exposure in Children," *Environmental Health Perspectives* 104, no. 3 (March 1996): 306-312.

Hua, Quan, Mike Barbetti, and Andrzej Z. Rakowski, "Atmospheric Radiocarbon for the Period 1950-2010," *Radiocarbon* 55, no. 4 (2013): 2059-2072.

Lopez, Adriana S., John Zhang, and Mona Marin, "Epidemiology of Varicella During the 2-Dose Varicella Vaccination Program—United States, 2005-2014," U.S. Department of Heath and Human Services *Morbidity and Mortality Weekly Report* 65, no. 34 (September 2016): 902-905.

Mahaffey, Kathryn R. et al., "National Estimates of Blood Lead Levels: United States, 1976-1980—Association with Selected Demographic and Socioeconomic Factors," *The New England Journal of Medicine* 307 (1982): 573-579.

Stamoulis, K. C. et al., "Strontium-90 Concentration Measurements in Human Bones and Teeth in Greece," *The Science of the Total Environment* 229 (1999): 165-182.

24. How to Win an Election

"3 Caseys Stirring Confusion," *Pittsburgh Post-Gazette*, October 21, 1976.

Full text of the polling questions collected by the Roper Center for Public Opinion Research:

(Do you think it is generally okay or not okay for people to use their cellphones in the following situations?)... At the movie theater or other places where others are usually quiet

Do you think sending a text message while driving, either on a cell phone or other electronic device, should be legal or illegal?

(Just off the top of your head, would you say you have a positive or negative image of each of the following.) How about... small business?

Do you think employers should or should not be able to obtain access to employees' genetic record, or DNA, without their permission?

As part of the effort to combat terrorism, would you support or oppose… creating criminal penalties for money-laundering involving terrorism?

Right now, a person needs to pass a test and get a license from the government in order to practice a number of occupations. Some people say this is necessary to guarantee that the public gets good services. Others say that it just increases the cost of the services. For each of the following, please tell me if you think government licensing is a good idea, or a bad idea?… Doctors

There has been a lot of discussion about what circumstances might justify the United States going to war again in the future. Do you feel if… the United States were invaded… it would be worth going to war again, or not?

Do you think the use of methamphetamines, sometimes known as 'crystal meth' should be made legal, or not?

Please tell me, are you satisfied or not satisfied with your… Friends

If a pill were available that made you twice as good looking as you are now, but only half as smart, would you take it, or not?

(Please tell me if you believe each of the following statements is true or false.)… When adults are supervising children in the water, they should be actively watching constantly, not reading or talking on the phone.

(Whether you are currently employed or not, think about people on the job, and tell me whether you think each of the following is okay or not okay.) Do you think it is okay, or not okay, to take more expensive things like computer or electronic equipment, telephones or other merchandise?

Some people say the following have become more common over the years, and I'd like your opinion about them. Do you think it is okay, or not okay to pay someone to do a term paper for you?

Let me read you some things some people have said they would like to see happen. Would you like to see a sharp decline in the number of people who suffer from hunger, or not?

(Let me read you some things some people have said they would like to see happen.) Would you like to see a decline in terrorism and violence, or not?

Let me read you some things some people have said they would like to see happen. Would you like to see an end to high unemployment, or not?

Let me read you some things some people have said they would like to see happen. Would you like to see… the elimination of starvation, or not?

(Let me read you some things some people have said they would like to see happen.) Would you like to see… a decline in prejudice … happen or not?

(Please tell me whether you believe that any of the following people or items can predict the future.)… The Magic Eight Ball

I'm going to read you a list of things people might say about the Olympic Games and I'd like you to tell me whether you personally agree or disagree with each statement. The statement is… the Games are great sports competition. (If necessary, ask:) Do you agree or disagree with that statement?

25. How to Decorate a Tree

"Airship Hangar in East Germany," *Nomadic-one*, August 18, 2011, http://www.nomadic-one.com/reflect /airship-hangar-east-germany.

"CNN/ORC Poll 12," conducted by ORC International, December 18-21, 2014.

Cohen, Michael P., *A Garden of Bristlecones* (Nevada: University of Nevada Press, 1998).

Foxhall, Emily, "Shopping Center Christmas Trees Compete for Needling Rights," *Los Angeles Times*, November 18, 2013, https://www.latimes.com/local/la -me-tree-20131119-story.html#axzz2lCOwKcfK.

Hall, Carl T., "Staying Alive/High in California's White Mountains Grows the Oldest Living Creature Ever Found," *SFGate*, August 23, 1998, https://www.sfgate .com/news/article/Staying-Alive-High-in-California-s -White-2995266.php.

Mahajan, Subhash, "Wood: Strength and Stiffness," in *Encyclopedia of Materials: Science and Technology* (Elsevier, 2001).

"Oldlist, A Database of Old Trees," Rocky Mounting Tree-Ring Research, accessed March 29, 2019, http://www .rmtrr.org/oldlist.htm.

Preston, Richard, "Tall for Its Age," *New Yorker*, October 9, 2006, https://www.newyorker.com /magazine/2006/10/09/tall-for-its-age.

Ray, Charles David, "Calculating the Green Weight of Wood Species," Penn State Extension, last modified June 30, 2014, https://extension.psu.edu /calculating-the-green-weight-of-wood-species.

Sussman, Rachel, *The Oldest Living Things in the World* (Chicago: University of Chicago Press, 2014).

26. How to Get Somewhere Fast

Chase, Scott et al., "The Relativistic Rocket," The Physics and Relativity FAQ, UC Riverside Department of Mathematics, last modified 2016, http://math.ucr.edu/home/baez/physics/index.html.

Davis, Tamara M. and Charles H. Lineweaver, "Expanding Confusion: Common Misconceptions of Cosmological Horizons and the Superluminal Expansion of the Universe," *Publications of the Astronomical Society of Australia* 21, no.1 (March 2013): 97-109.

"Plot of Distance (in Giga Light-Years) vs. Redshift According to the Lambda-CDM Model," Wikimedia Commons, accessed March 29, 2019, https://en.wikipedia.org/wiki/Redshift#/media/File:Distance_compared_to_z.png.

27. How to Be on Time

15 "U.S. Code § 262. Duty to Observe Standard Time of Zones," *Code of Federal Regulations,* Mar. 19, 1918, ch. 24, § 2, 40 Stat. 451; Pub. L. 89-387, § 4(b), Apr. 13, 1966, 80 Stat. 108; Pub. L. 97-449, § 2(c), Jan. 12, 1983, 96 Stat. 2439.

49 "CFR Part 71—Standard Time Zone Boundaries," *Code of Federal Regulations,* Secs. 1-4, 40 Stat. 450, as amended; sec. 1, 41 Stat. 1446, as amended; secs. 2-7, 80 Stat. 107, as amended; 100 Stat. 764; Act of Mar. 19, 1918, as amended by the Uniform Time Act of 1966 and Pub. L. 97-449, 15 U.S.C. 260-267; Pub. L. 99-359; Pub. L. 106-564, 15 U.S.C. 263, 114 Stat. 2811; 49 CFR 1.59(a).

Allen, Steve, "Plots of Deltas between Time Scales," UC Observatories, accessed May 20, 2019, https://www.ucolick.org/~sla/leapsecs/deltat.html.

Morrison, L.V. and F.R. Stephenson, "Historical Values of the Earth's Clock Error ΔT and the Calculation of Eclipses," *Journal for the History of Astronomy* 35, no. 120 (2004): 327-336.

Na, Sung-Ho, "Tidal Evolution of Lunar Orbit and Earth Rotation," *Journal of the Korean Astronomical Society* 47, no. 1 (April 2012): 49-57.

Nazarli, Amina, "Azerbaijan Cancels Daylight Saving Time—Update," *Azernews,* March 17, 2016, https://www.azernews.az/nation/94137.html.

28. How to Dispose of This Book

Caporuscio, Florie et al., "Salado Flow Conceptual Models Final Peer Review Report," Waste Isolation Pilot Plant, U.S. Department of Energy, March 2003.

"The Deterioration and Preservation of Paper: Some Essential Facts," Library of Congress, accessed May 3, 2019, https://www.loc.gov/preservation/care/deterioratebrochure.html.

Erdincler, Aysen Ucuncu, "Energy Recovery from Mixed Waste Paper," *Waste Management and Research* 11, no. 6 (November 1993): 507-513.

Jackson, C.P. et al., "Sealing Deep Site Investigation Boreholes: Phase 1 Report," Nuclear Decommissioning Authority, May 14, 2014.

Jefferies, Nick et al., "Sealing Deep Site Investigation Boreholes: Phase 2 Report," Nuclear Decommissioning Authority, March 23, 2018.

Pusch, Roland and Gunnar Ramqvist, "Borehole Project-Final Report of Phase 3," Swedish Nuclear Fuel and Waste Management Co, 2007.

Pusch, Roland and Gunnar Ramqvist, "Borehole Sealing, Preparative Steps, Design and Function of Plugs—Basic Concept." *SKB Int. Progr. Rep. IPR-04-57* (2004).

Pusch, Roland et al., "Sealing of Investigation Boreholes, Phase 4-Final Report," Swedish Nuclear Fuel and Waste Management Co., 2011.

Sequeira, Sílvia Oliveira, "Fungal Biodeterioration of Paper: Development of Safer and Accessible Conservation Treatments" (PhD diss., NOVA University Lisbon, 2016).

Teijgeler, René, "Preservation of Archives in Tropical Climates: An Annotated Bibliography," International Council on Archives (Jakarta, 2001).

Ximenes, Fabiano, "The Decomposition of Paper Products in Landfills," Appita Annual Conference (2010): 237-242.

Index

How to
Change a Light Bulb